Benchmarking Industry-Science Relationships

OECD

ORGANISATION FOR ECONOMIC CO-OPERATION AND DEVELOPMENT

ORGANISATION FOR ECONOMIC CO-OPERATION
AND DEVELOPMENT

Pursuant to Article 1 of the Convention signed in Paris on 14th December 1960, and which came into force on 30th September 1961, the Organisation for Economic Co-operation and Development (OECD) shall promote policies designed:

- to achieve the highest sustainable economic growth and employment and a rising standard of living in Member countries, while maintaining financial stability, and thus to contribute to the development of the world economy;

- to contribute to sound economic expansion in Member as well as non-member countries in the process of economic development; and

- to contribute to the expansion of world trade on a multilateral, non-discriminatory basis in accordance with international obligations.

The original Member countries of the OECD are Austria, Belgium, Canada, Denmark, France, Germany, Greece, Iceland, Ireland, Italy, Luxembourg, the Netherlands, Norway, Portugal, Spain, Sweden, Switzerland, Turkey, the United Kingdom and the United States. The following countries became Members subsequently through accession at the dates indicated hereafter: Japan (28th April 1964), Finland (28th January 1969), Australia (7th June 1971), New Zealand (29th May 1973), Mexico (18th May 1994), the Czech Republic (21st December 1995), Hungary (7th May 1996), Poland (22nd November 1996), Korea (12th December 1996) and the Slovak Republic (14th December 2000). The Commission of the European Communities takes part in the work of the OECD (Article 13 of the OECD Convention).

Publié en français sous le titre :
LES RELATIONS INDUSTRIE-SCIENCE
Une évaluation comparative

FOREWORD

This report analyses the changing role of industry-science relationships (ISR) in national innovation systems, proposes a conceptual framework for their assessment, presents indicators on international differences in ISR configuration and intensity, and identifies good practices for their improvement. It also applies the benchmarking framework for an in-depth comparison of ISR in France and the United Kingdom, and takes a special look at industry-science relations in Japan.

The report summarises the findings of a two-year OECD project carried out under the guidance of the OECD's Committee for Scientific and Technology Policy and of its Working Party on Technology and Innovation Policy. It benefited from information provided by Member countries and the results from a major international Conference in Berlin that was jointly organised by the German Ministry for Research and Education (BMBF) and the OECD. The pilot comparative study on France and the United Kingdom was made possible thanks to John Barber, UK Department of Trade and Industry, and Pierre-Yves Mauguen, French Ministry of Research.

Jean Guinet, with support from Mario Cervantès, managed the overall project, drafted Part I and edited the publication, which also benefited from contributions by Benedicte Callan and Yuki Fukasaku. The case studies on France and the United Kingdom were drafted by Rémi Barré, and John Rigby and Luke Georghiou, respectively. Akira Goto and Ryuji Shimoda prepared the chapter on Japan.

The report is published on the responsibility of the Secretary-General of the OECD.

TABLE OF CONTENTS

SUMMARY

- In a knowledge economy, science is exerting a more important and direct influence on innovation, especially in fast-growing new industries. The intensity and quality of industry-science relationships (ISRs) thus play an increasing role in determining returns on investment in research, in terms of competitiveness, growth, job creation and quality of life. They also determine the ability of countries to attract or retain increasingly mobile qualified labour.

- ISRs are undergoing fundamental changes prompted by globalisation and other factors as part of an overall trend towards accelerated development of a market for knowledge. The most visible transformations are the emergence of broad alliances between universities and firms, and growing activity in the realm of commercialisation of results through licensing of intellectual property and spin-off companies (*e.g.* during the 1990s, US universities more than doubled their propensity to patent).

- Whereas the experiences of some countries, especially the United States, suggest that research and commercialisation goals are not only compatible but can reinforce one another, there is accumulating evidence that many OECD countries are lagging behind in the modernisation of ISRs. At the same time, countries that are forging ahead in building new, more flexible models of ISRs, are experiencing new challenges.

- Generally, governments lack information and tools to monitor ISRs, evaluate their efficiency and learn from each other in the search for good policy practices.

Benchmarking ISRs

- To benchmark ISRs is to compare their relative efficiency in meeting and reconciling the needs of the main stakeholders (governments, industry, public research organisations, civil society), and to relate differences in performance to observable characteristics of industry-science linkages, focusing on those aspects that are amenable to policy. To this end, industry-science linkages should be characterised and evaluated along three dimensions: nature and relative importance of the channels of interaction; their incentive structures; and their institutional arrangements.

- Regarding the channels of interaction, the report focuses on labour mobility and spin-offs. Despite a general trend towards relaxing regulatory constraints, the low rate of mobility of researchers between the private and public sectors remains in many countries a major bottleneck to ISRs. The contribution of spin-offs from publicly funded research to innovation is significant, especially in the information technology and, increasingly, the biotechnology/medical technologies sectors. The rate of spin-off formation, which may be a good indicator of the health of ISRs, is about three to four times higher in North America than in most other OECD countries.

- Regarding incentive structures, the report focuses on intellectual property rights and on research evaluation systems. In nearly all OECD countries there has been a trend towards transferring ownership of publicly funded research results from the state (government) to the (public or private) agent performing the research. Where countries differ is in the allocation of ownership among performing agents (research institution *vs.* individual researcher).

- Regarding institutional arrangements, the report views university-based systems of ISRs as enjoying a comparative advantage when science-based innovation increasingly requires multidisciplinarity and builds on people-based interactions. In the last decade, a majority of OECD countries have redirected public R&D investment towards universities to the detriment of research institutes. However, this shift has not improved ISRs to the same extent everywhere. A major reason is that decentralised university systems, in which universities enjoy more freedom in their research policy and relations with industry, are more responsive to opportunities for ISRs than are centralised ones. The report assesses the strengths and weaknesses of different models for organising commercialisation activities in the public research sector, and demonstrates that improving the contribution of public research institutes to innovation often requires updating their missions and improving their linkages with universities.

- There is no single model for commercialising public research that could be used for evaluating the relative performance of national systems of ISRs. As this performance should be measured by taking into account different dimensions (and hence indicators), a number of countries appear to perform well with respect to several indicators. Moreover even in countries where performance measures would indicate a low level of ISRs, good practice examples can be found for improving the structure of and framework conditions for ISRs.

- Mutual learning from good practices requires continuous and intensified efforts to monitor and assess ISRs from an international perspective. However, in order to effectively inform the policy decision-making process, benchmarking should not be carried out as a "beauty contest", based on crude indicators, but rather should be organised as a learning process through which all major stakeholders in ISRs find new inspirations and motivations for improving their contribution to the innovation system.

- A pilot study comparing ISRs in France and the United Kingdom was undertaken to develop and test a conceptual framework and a methodology for such meaningful and policy relevant benchmarking. Its results are reported in this book. One of its important findings is that social networks, as shaped by the education system, exert a strong influence on the national patterns of ISRs. Another is the importance of avoiding generalisation and the need to distinguish between three types of ISRs: those involving multinational enterprises and world class universities; relations between universities and high–technology small firms; and relations developing in a regional context between firms (often SMEs looking for shorter-term problem-solving capabilities) and the local university.

Policy conclusions

The main policy implications of the above findings were discussed in the German-OECD Conference on ISRs in Berlin,[1] and in relevant OECD forums, including in the context of the OECD Growth project. It was agreed that, while publicly funded research organisations (universities and public laboratories) and industry are best placed to determine how their collaboration can be enhanced in practice, governments have the responsibility for setting the basic rules and institutional frameworks that reflect the public interest but provide the right incentives to firms, public researchers and organisations alike. In all countries, policy action is especially important in six areas:

- *Giving greater priority to basic and long-term mission-oriented research in government S&T programmes.* Basic and long-term research – whether motivated by scientific curiosity or by the challenges facing industry and society – produce new scientific and technical knowledge that is increasingly important in driving innovation. Changes in business R&D strategies are generally accentuating longstanding disincentives for private industry to invest in fundamental research, thus heightening the need for government support.

- *Ensuring appropriate frameworks for intellectual property rights.* Governments must establish clear rules and guidelines with regard to the intellectual property resulting from publicly funded research, while granting sufficient autonomy to research institutions. A good practice is to grant intellectual property rights to the performing research organisation while ensuring that individual researchers or research teams can share in the rewards. An interim conclusion is that a good practice might be to grant IPR ownership to the performing research organisation but to ensure that researchers enjoy a fair share of the resulting royalties. Globalisation of research accentuates the need for additional efforts to harmonise IPR regimes and practices at international level. Currently, far too much time is wasted in attempting to work out the details and differences in the patenting and licensing policies of different countries.

- *Matching supply and demand of scientific knowledge.* Regulatory reforms related to IPRs and the licensing of publicly funded research should be complemented by measures (such as the establishment of technology licensing offices, public/private partnerships in funding R&D, stimuli for co-operation with business, and support for spin-off formation) that stimulate business demand for scientific inputs and improve the ability of public research organisations to transfer knowledge and technology to the private sector.

- *Improving the governance of universities and public laboratories.* Public laboratories can be made more responsive to emerging needs by establishing new mechanisms for priority-setting and funding that reflect industry input and tie funding to performance, as well as by strengthening their links with the training and education system. Additional efforts to break down disciplinary boundaries will enable them to better engage in emerging scientific and technical areas. In many countries, universities would benefit from greater autonomy in decision making coupled with more programmatic R&D funding. Institutional support remains important but more competitive funding instruments are needed to improve the quality of research results while ensuring that fields of science of high economic importance receive attention.

- *Safeguarding public knowledge.* Setting clear rules on IPRs is key but not sufficient to achieve a balance between commercial aims and the research and teaching missions of the public research institutions. Governments must ensure sufficient public access to knowledge from publicly funded research. It must also acknowledge the risks to the research and innovation system that may result if the IP protection granted is too strong and non-exclusive licensing too rare. Finally, ethical guidelines for and by public research institutions are necessary to prevent or resolve conflicts of interest among the institutions and researchers involved in collaboration with industry.

- *Promoting the participation of smaller firms.* Young technology-based firms play a key role in linking science to markets. Governments rightly attach priority to encouraging spin-offs from public research to stimulate innovation. Spin-offs fill a gap between research results and innovative products and services. They are also a means for universities to broadly license technology. However, there is also a case for public support and incentives to existing SMEs and especially those in mature industries in order to help them link up with the science base and enhance innovation capacity.

- *Attracting, retaining and mobilising human resources.* Strong demand for highly skilled personnel increasingly extends across borders, raising concerns about a "brain drain" in some countries in which the loss of one or two key individuals can undermine research capabilities. For companies and research institutions, keeping talent requires investments in in-house training, career growth potential as well as excellent research working conditions. To attract students at university, graduate programmes must better integrate interdisciplinarity and contacts with industry in training and research. For governments, removing barriers and disincentives to mobility and flexibility in research employment is also essential. Worker mobility is a critical element of industry-science relations and can be enhanced by regulatory reforms that allow public researchers to work more closely with private industry.

- *Improving the evaluation of research.* Evaluation of publicly funded research must evolve in response to the considerable expansion of the commercialisation activities of universities and public research institutes, and evaluation criteria must take into account that excellence in research and training of graduates has become, at least in some disciplines, more tied to applications in industry. Evaluation criteria need to recognise the quality of the research, its potential social and economic impact, and the value of university research in educating students. In this area, national initiatives should be complemented by further efforts at international level to develop benchmarking indicators and methodologies, and promote the use of foreign expertise in national evaluation.

- *Responding to globalisation.* The accelerating internationalisation of large firms' R&D activities, as well as increased global competition to attract entrepreneurs, research talent and venture capital, challenge national policies to promote industry-science relationships. On the one hand, the participation of foreign firms in national programmes is increasingly key to their success. On the other, national research institutes and universities must be encouraged to internationalise their linkages with industry.

- *Building on existing innovative networks and clusters.* The most successful industry-science partnerships involve links between publicly financed research organisations and a cluster of local industries. Governments should accept the fact that giving more weight to commercialisation objectives in managing the science system, including the allocation of core funding, is likely to accentuate the polarisation of university research capabilities around existing centres of excellence. The promotion of industry-science relationships should be an integral part of an overall cluster- and network-based innovation policy strategy.

NOTE

1. The Proceedings of the Berlin Conference, prepared by the OECD Secretariat and the German Ministry of Research and Education (BMBF), can be found on the OECD Web site at www.oecd.org/sti/innovation.

Part I

BENCHMARKING INDUSTRY-SCIENCE RELATIONSHIPS

RATIONALE, METHODOLOGY AND RESULTS

Chapter 1

THE GROWING AND CHANGING ROLE OF INDUSTRY-SCIENCE RELATIONSHIPS IN INNOVATION-LED GROWTH

Challenges and issues

"The nation that fosters an infrastructure of linkages among and between firms, universities and government gains competitive advance through quicker information diffusion and product deployment" (US Council on Competitiveness, 1998). In other words, today, the performance of an innovation system increasingly depends on the intensity and effectiveness of the interactions between the main actors involved in the generation and diffusion of knowledge. The debate on the "new economy" has led to a wider recognition of both the increasing role of innovation as a determinant of growth and the changing nature of innovation processes (OECD, 2000, 2001a, 2001b). It also points to the vital role that healthy and adaptive industry-science relationships (ISRs) play in the development of fast-growing new industries and in training, retaining and attracting highly qualified labour (Box 1).

As a result, science-industry linkages have grown in importance as a central concern for government policy in recent years. This interest coincides with a number of new developments in the nature of ISRs, such as the emergence of broad alliances between universities and firms, and growing activity in the realm of commercialisation of results through licensing of intellectual property and spin-off companies. The intensification and diversification of industry-science relationships is most notable and well documented in the United States (Figure 1), but can also be observed in other countries, including those where informal (hardly measurable) mechanisms of interaction have traditionally played a greater role, such as France or Japan (see country case studies in Part II). This signals deeper ongoing transformations in the respective role of and forms of co-operation/competition between curiosity-driven scientific research, mission-oriented public research and profit-driven business R&D, due to the combined effect of the following factors:

- Technical progress accelerates and markets expand exponentially in areas where innovation is directly rooted in science (biotechnology, information technology, but also new materials).

- New information technologies allow easier and cheaper exchange of information between researchers.

- Industry demand for linkages with the science base increases more broadly, as innovation requires more external and multidisciplinary knowledge, tighter corporate governance leads to the downsizing and shorter-term orientation of corporate labs,[1] and more intense competition forces firms to save on R&D costs while seeking privileged and rapid access to new knowledge.

Links to the science base are more important than in the past

Academic work is becoming increasingly important for industrial activities. Basic scientific research is the source of many of the technologies that are transforming society, including the Internet. Innovation in key sectors such as information technology and biotechnology, in particular, is closely linked to advances in basic science. In addition, publicly funded research provides the skilled graduates that are essential to firms wishing to adopt new technologies, new instruments and methods for industrial research and an increased capacity for problem solving. Scientific institutions also play a role in the formation of the world's research and innovation networks, which are now considered crucial to successful technology diffusion and innovation (OECD, 2000, 2001b).

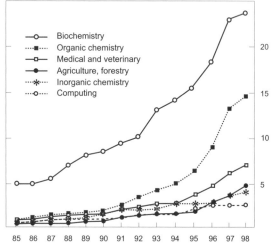

Figure b1. Science linkages in selected countries

Measured by the average number of scientific papers cited in US patents depending on applicants' nationality

- United States
- United Kingdom
- France
- Germany
- Japan

Source: NISTEP (2000), based on data from CHI Research Inc.

Figure b2. Science linkages in selected areas in the United States

Measured by the average number of scientific papers cited in US patents for US applicants

- Biochemistry
- Organic chemistry
- Medical and veterinary
- Agriculture, forestry
- Inorganic chemistry
- Computing

Source: NISTEP (2000), based on data from CHI Research Inc.

The science system's influence on innovation is now more direct

Available evidence suggests that in many fields technological innovation makes more intensive use of scientific knowledge. In the United States, almost three-quarters of the references to scientific publications listed in US patents are from public science (Narin, 1997). The number of references to public science nearly tripled over the six-year period covered. Between 5% and one-third of new products, depending on the sector, could not have been introduced without direct input from recent academic research; in addition, the time delay from academic research to industrial practice has in average shortened from seven years to six during the 1990s (Mansfield, 1998).

Table b1. Citations from national sources in US-issued patents, 1990-97

Share of national scientific sources	21.3	18.4	15.7	13.9	18.9	24.4	63.7
Citation ratio*	10.4	4.3	24.2	7.0	11.5	3.2	1.8
Country of citing patent inventor	Australia	Canada	Finland	Netherlands	Sweden	United Kingdom	United States

* Share of national scientific sources divided by share of the country in world publications.
Source: Narin *et al.* (2000).

Science-industry linkages are unevenly developed accross OECD countries

Table b1 shows that linkages between technology and science have a strong national component, even for small countries. Figure b1 shows that the intensification of science-industry relationships, as reflected in patents, is occuring in all countries but at a strinkingly variable pace.

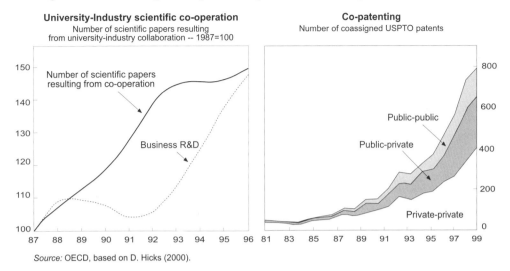

Figure 1. **The increasing intensity of industry-science relationships in the United States**

University-Industry scientific co-operation
Number of scientific papers resulting
from university-industry collaboration -- 1987=100

Co-patenting
Number of coassigned USPTO patents

Source: OECD, based on D. Hicks (2000).

- The ability to respond to new social needs (health of ageing populations, environmental sustainability, security) calls for innovations that often require mobilising complementary competencies of the public and private research sectors (Box 2).

- Financial, regulatory and organisational changes boost the development of a market for knowledge by enabling the financing and management of a wider range of commercialisation activities.

- Restrictions on core public financing (Table 1) encourage universities and other publicly funded research organisations to enter this booming market, especially when they can build on existing solid linkages with industry.

These driving forces operate more forcefully in some countries than in others and they do not encounter the same obstacles in all national contexts. The main concern of most OECD countries is that they could lag behind in the modernisation of ISRs. At the same time, countries that are forging ahead in building a new, more fluid, model of ISRs experience novel problems in fine-tuning it (Box 3). These problems need to be taken into account by other countries when designing their approach to catch up with best practices.

Case studies and anecdotal evidence on success factors in managing science-industry relationships are accumulating, but governments find it difficult to relate these experiences to their own concerns when assessing the situation from a nation-wide perspective, evaluating recent reforms and determining the need for further policy initiatives. Evaluation of industry-science linkages is generally carried out at the level of specific research institutions or public support programmes, providing results that are difficult to compare from one institution to the other and internationally. Identification of and learning from good practices in joint creation, transfer and sharing of knowledge between the public and private research sectors are impaired by the lack of agreed methodologies and indicators to measure performance.

Table 1. **Higher education expenditures on R&D by funding source for seven European countries, 1983-97**

Percentages

	Total public share	General university funds (GUF)	Direct government funds	Foreign	Business	Other income	Private non-profit organisations
1983	94.0	68.3	25.7	0.6	2.9	1.1	1.5
1991	89.4	61.7	27.7	1.6	5.5	1.2	2.3
1995	85.6	59.0	26.6	3.2	5.7	1.8	3.7
1997	84.6	57.9	26.8	3.5	6.4	1.7	3.8

Note: Denmark, France, Germany, Italy, Ireland, the Netherlands, the United Kingdom; figures represent the weighted average.
Source: OECD.

The first objective of a benchmarking exercise is therefore to characterise the current state and evaluate the direction and pace of evolution of the various national ISRs, with a view to helping governments determine the need and scope for improvement and subsequently monitor progress:

- What are the potential benefits of ISRs for the different stakeholders? What are the respective roles of the different channels of industry-science relations in realising this potential? Are some of these channels becoming more important with the emergence of a knowledge-based economy?

- How to assess whether and in which respects a national system of ISRs is keeping pace with evolving best practices? Which indicators should be used?

Box 3. New issues in managing ISRs: the example of structural genomics*

Structural genomics is a new and exciting field of basic research at the intersection of the public and private domains. It follows up the dramatic advances in DNA sequencing. There now exist complete, publicly available genomic maps for many organisms, including the cholera bacterium, the all-purpose weed *Arabidopsis thaliana*, the worm *Caenorhabditis elegans*, the fruit fly, and that well-known primate, *Homo sapiens*. Sequences can be useful in themselves (they are the basis, for example, of the study of the genetic dimension of diseases), but the acquisition of this knowledge is merely the beginning of a much wider quest to determine the functions of the proteins whose composition is encoded by DNA. Knowledge of the three-dimensional physical structures of these proteins is a crucial part of the pursuit, since structure is closely linked to biological function. The availability of complete genomes has inspired some scientists to propose that the corresponding structural information should be obtained for hundreds of thousands of proteins, leading to a potential quantum leap in the understanding of the totality of the integrated functions of organisms. This bold concept has been termed "structural genomics".

Structural genomics initiatives have arisen almost simultaneously in the academic and industrial communities. Governmental decisions are influenced by an awareness that post-genomics may be a major source of a new generation of drugs and therapies and of future industrial competitiveness and wealth creation. Governments need to decide what are the most appropriate relative roles and relationships of the public and private sectors, and how public money should be spent to optimise desirable outcomes: advancing science, promoting industrial growth, and delivering the fruits of research to the public. Some public authorities are concerned about being left behind. US researchers, benefiting from generous government programmes, have built up a strong lead in the field, with their Japanese colleagues not far behind, while the Europeans are searching for ways to advance via the right mix of national and EU programmes. Three aspects of structural genomics deserve the special attention of science policy makers: infrastructures (facilities and equipment); the scope of the research; and intellectual property rights (IPRs).

Research infrastructures. DNA sequencing has become highly automated and the acquisition of new knowledge is a function of the financial investment in people (mostly technicians) and relatively inexpensive, commercially available machines. Structural analysis, on the other hand, consists of a series of complex and difficult steps, each one of which requires specialised expertise and the availability of the appropriate apparatus and resources. The work must be done by a PhD-level scientist, using sophisticated experimental schemes and instrumentation, high-performance computers and specialised software. Thus, government officials have to decide on the extent to which they wish to invest public funds to build large research infrastructures for use by private companies. During the next few years, policy makers and laboratory officials may be faced with difficult choices about allocation of X-ray sources among paying and non-paying users. Public funds may be needed for very expensive new facilities such as "free electron lasers" which may allow scientists to analyse entire new classes of proteins.

The scope of structural genomics projects. Because structural analysis is so expensive and time-consuming, great care must be taken in choosing the right set of proteins for any structural genomics project. Industrial researchers are more likely to focus on molecules that are linked to diseases (for example, viral enzymes) since these may be promising "drug targets". Academic researchers may be more inclined to study proteins that provide insight into broader questions, for example, selected pathways of cellular metabolism, or evolutionary theory. Since there is no clear dividing line between these lines of inquiry, some co-ordinating mechanism probably needs to be established to promote exchange of information about which proteins are being analysed, and to avoid unnecessary duplication of effort.

.../...

The second objective is to help policy makers determine what should be improved and how, through a comparative assessment of country experiences in addressing the following main issues:

- What are the most important bottlenecks in ISRs: low demand from the private sector, low quality or share of publicly funded industry-relevant research, obstacles to researchers' mobility, inability to manage contractual relations (licensing, research contract), ineffective intermediaries, lack of entrepreneurship in the research community, weakness of social networks and international linkages, etc.?

- Which levers are available to promote the desired changes (*e.g.* financial incentives, regulatory reform, organisational change, new mechanisms and criteria to allocate funding or to evaluate public research and researchers), and what are their comparative strengths, limitations and political feasibility?

- Are more intensive ISRs always more effective? How far should universities and public labs be allowed or pushed to develop their commercialisation activities? How to cope with the risk of crowding out private initiative and distorting the market for technological services?

- What safeguards should be in place to ensure that publicly funded research institutions do not strengthen their linkages with industry at the expense of their main missions (generation and diffusion of knowledge through free research and education, mission-oriented research to serve public interest, impartial scientific expertise)? In particular, how to ensure that increased patenting and industry involvement in areas close to basic research do not disrupt scientific work or weaken public confidence in science? Should governments place limits on the possibilities of universities to accept publication restrictions when co-operating with industry?

- To what extent should a policy to promote ISRs be implemented through a national set of rules and incentives, as opposed to broader guidelines for decentralised experiments at the level of regions or individual research institutions? In particular, should governments seek harmonisation of IPR practices in all publicly funded research organisations?

Trends in ISRs

To benchmark ISRs is to compare their relative efficiency in meeting and reconciling the expectations of the major stakeholders and to relate differences in performance to observable characteristics of industry-science linkages, focusing on those aspects that are amenable to policy. It is therefore important to clarify what the stakeholders' expectations really are, whether they are changing in the knowledge economy, and whether this increases the relative importance of certain channels, incentives or institutional arrangements, reduces conflicts of interest in the innovation system or creates new ones.

Changing stakeholders' objectives and needs

In theory, *governments* should expect that efficient ISRs reduce systemic failures in economy-wide knowledge generation and diffusion, thereby increase the social return on public investment in research and ultimately contribute to increased productivity and growth. However, their real objectives are less abstract, stable and consistent, being influenced by the economic cycle (notably the labour market situation), the evolving priorities of technology and innovation policy, and the most pressing issues in the management of the science system (*e.g.* employment of PhDs, shortage of finance). In the last decade, universities in many countries have been called upon to compensate for the decline of public research institutes in the commercialisation of public research. In addition, in the recent period, policy attention in most OECD countries has tended to focus increasingly on the role of ISRs in fostering entrepreneurship in fast-growing new industries, often to the neglect of other important contributions of the science system.

Publicly funded research organisations value relationships with industry for different reasons depending on their main mission. *Universities* cultivate industry contacts to ensure good employment prospects for students, keep curricula up-to-date in some disciplines and obtain financial or in-kind support to reinforce and expand their research capabilities beyond what would be allowed by core funding. Leading research universities now adopt more ambitious goals, including strategic alliances with firms to consolidate their position in innovation networks and ensure they get their share of the booming market for knowledge. Smaller universities are tempted to transform part of their research departments into business support units and contract research organisations, especially in countries, such as the United Kingdom, which have imposed tough competition for core funding. *Large multidisciplinary public research institutes* have always had close links with the private sector in areas where industry is an important player in the whole research spectrum, including fundamental research. They now need to adapt their interface with industry to the requirements of new science-based industries where start-ups and small firms are important players. *Mission-oriented public research institutes* have developed almost organic linkages with the part of industry that offers complementary competencies in responding to government procurement. The need to diversify their activities away from stagnant or declining core activities is driving the ongoing changes in their relations with industry.

Innovation surveys demonstrate that improved access to better trained human resources is by far the main benefit that *industry* expects from linkages with publicly financed research, and this is not likely to change in the future given the risk of persistent shortages of highly qualified labour. Among other benefits (that include also networking and clustering opportunities or access to problem-solving capabilities), privileged access to new scientific knowledge is taking on a new importance. Whereas industry remains a significant actor of the science system, especially in chemistry, physics and basic engineering (NSF, 1998), it relies increasingly on public research to complement its own growing R&D efforts. However, industry views diverge concerning the preferred channels of access to publicly funded research. For example, increased patenting by publicly funded organisations yields more benefits to small firms than to large ones that have long-established close links with public research. In the service sector, many firms see the increased commercialisation activities of universities as unfair competition, while others have created a business out of assisting this process.

The importance of informal and human resource-related linkages

Formal mechanisms of ISRs are only the tip of the iceberg (Figure 2). The bulk of industry-science relations take place through informal and indirect channels and also through unrecorded direct channels in countries where the regulatory framework has been fairly restrictive in the past. In the United Kingdom, innovation surveys show that, while almost half of manufacturing firms consider universities to be an important source of innovation, only 10% have developed formal relationships with them (SPRU, 2000). As mentioned above, the flow of skilled personnel to industry is the single most important channel of ISRs. Informal networks between faculty and former graduates and between former public researchers and their lab of origin account for a large, although difficult to measure,[2] share of the total amount of knowledge exchanged between industry and public research. New information and communication technologies can only reinforce the role of these social networks in ISRs. By focusing on what is measurable through conventional techniques, economists and governments generally underestimate these human resource-related linkages. They tend to overlook the fact that access to scarce human resources is always a key objective of industry in considering the merits of any type of linkage, formal or not, with public science.

Increased commercialisation of public research

This is not to deny the importance of formalised linkages, especially contract research, and the fact that the most spectacular ongoing change in ISRs is the accelerating development of some of these formal relationships, especially spin-offs (see section below) and patents.

The large increase in patents filed by the private sector, public research, or jointly by companies and public research, underscores the growing transformation of knowledge into an economic asset. In the United States, university patenting has increased more rapidly than university research spending and more rapidly than national patenting rates. US universities more than doubled their propensity to patent during the 1990s, as did the US public laboratories, starting from a lower level (Figure 3).

Figure 2. **Formal mechanisms of ISRs: the tip of an iceberg**

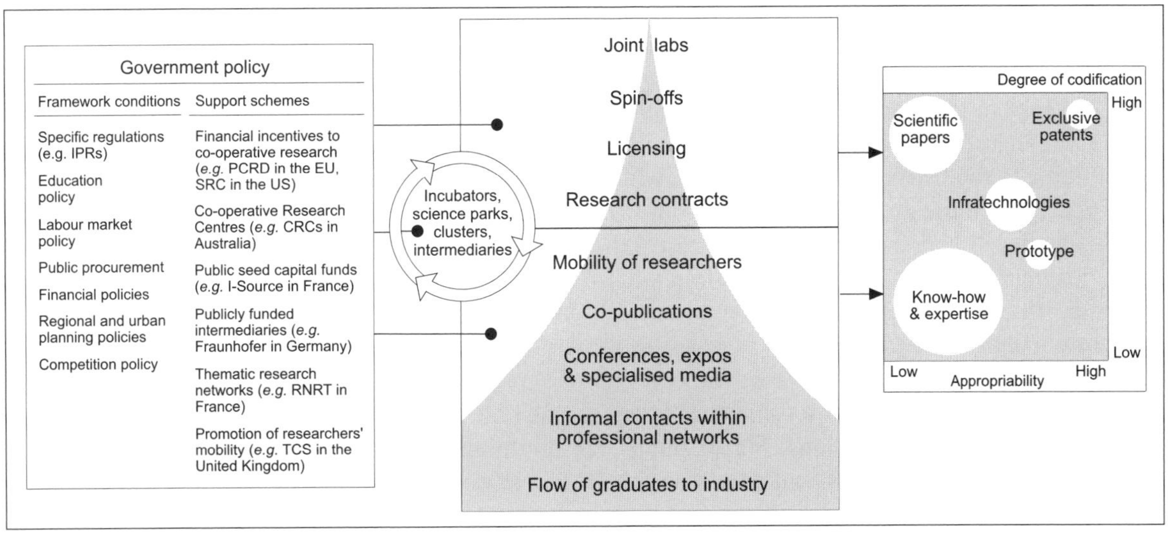

Source: OECD.

The lack of comparable data makes international comparisons difficult. However, anecdotal evidence suggests that US public research leads, but is not alone in the "patenting race". For example, university patenting in Germany appears to be quite prolific; the share of patent applications listing university professors as inventors has been rising steadily since the 1980s, and represented 4% of total applications by the mid-1990s (BMBF, 1997; Schmoch *et al.*, 2000). In Australia, the largest public research organisation (CSIRO) appears to lag behind top US research universities but performs better than the average US university in terms of royalty revenues as a percentage of R&D spending (Thorburn, 1999). In the first half of the 1990s, licensing income of four of the main French labs (CNRS, INSERM, INRA, INRIA) was equivalent to only 0.6% of their budgets, *i.e.* less than one-tenth the licensing revenues of US universities, although this percentage has increased rapidly since.

Japanese public research has distinctively weaker, although steadily growing (Table 2), patenting activities than its homologues in other advanced OECD countries (with a yearly average of around 150 patents from universities in the 1990s, *i.e.* less than half the invention disclosures by publicly funded organisations in Massachusetts alone). Another indicator is the negligible share of the universities in total patents in Japan – less than 0.1% compared to about 3% for US universities (Hashimoto, 1998; Howells, 1998).

Table 2. **Number of patent applications by Japanese national universities**

	1994	1995	1996	1997	1998	1999
Domestic	26	25	35	75	138	191
Foreign	21	23	22	34	73	93

Source: Shimoda and Goto (See Chapter 6, Part II).

However, in considering the policy implications of the recent surge in patenting, using the United States as a benchmark, it is important to keep in mind the following facts and emerging issues:

- *Revenues from patenting do not significantly reduce the need for other sources of funding, except in rare cases.[3]* In the United States, gross revenues from licences represent on average less than 3% of R&D funding of US universities and less than 2% of R&D spending of public labs. In the University of California, which tops the list of US universities in terms of licensing income, these revenues represent only 6% of total federal research. Net revenues are much smaller – and often negative – given the high and escalating cost of managing IPRs. For example, in 1997-98 the Australian CSIRO spent AUD 4.7 million for legal and patent portfolio management costs compared with AUD 5.26 million of income from patents.

- *Patenting is not a reliable indicator of scientific output.* The distribution of academic patents is highly skewed towards bio-medical sciences; the bulk of revenues from patenting come from a few successful inventions. For example, the doubling of income from patents by the French CNRS in 1997 was attributable in large part to a single product, Taxoter, which accounted for 40% of total licensing income in that year.

- *Government played a crucial role in spurring patenting activities but was aided by other changes.* Changes in the intellectual property regime (Bayh-Dole Act) was among the key factors behind the rise of US university patenting and licensing in the last two decades. However, the new regime built on a longstanding tradition of industry-university collaboration facilitated by the autonomous status of the research universities (Mowery, 1998). Other factors also played a role: institutional change (proliferation of technology and transfer offices, partly in reaction to Bayh-Dole), technological evolution (rise of the biotechnology and IT industries), financial incentives (reduction in government funding).

- *The main contribution to innovation of increased patenting is not to make public sector research more commercially relevant but to improve information on the existence and location of commercially relevant research results* (Henderson *et al.*, 1998).

- *Buoyant patenting activities should not overshadow the parallel development of other forms of ISRs.* University-industry research centres (UIRCs) in the United States, and similar mechanisms in other countries (*e.g.* CRCs in Australia), have become popular mechanisms for fostering public-private co-operation and are successful at both leveraging government support for academic research and orienting the latter towards more applied problems.

- *Greater autonomy of publicly funded research organisations increases their contribution to innovation through patenting and other means when it is paralleled by greater accountability.* Centralised systems with restrictive regulatory frameworks but low accountability reduce the responsiveness of public research to industrial needs and encourage the development of "grey" relations that would be prohibited in the name of public interest in more "liberal" and decentralised systems.

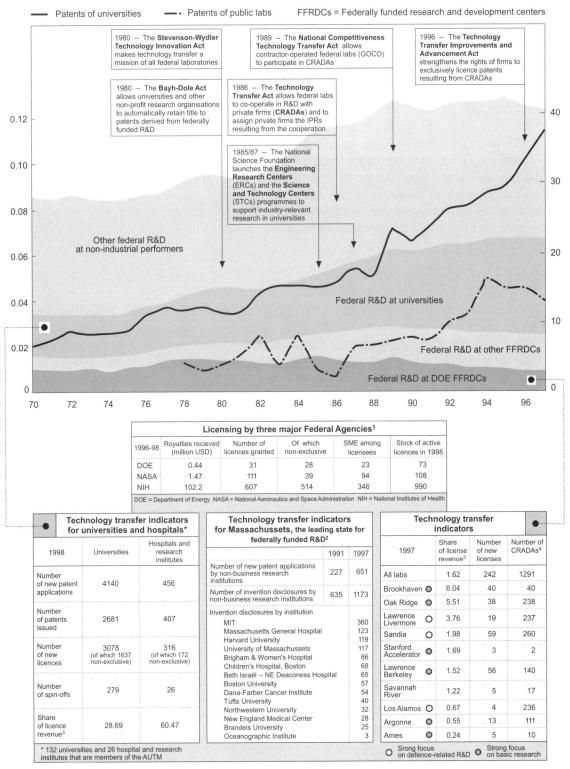

Figure 3. **Publicly funded patents per million USD of research expenditure** (left scale) **and federally funded R&D by non-industrial performers** (Right scale; billions of 1995 USD)

— Patents of universities   —·— Patents of public labs   FFRDCs = Federally funded research and development centers

1980 -- The **Stevenson-Wydler Technology Innovation Act** makes technology transfer a mission of all federal laboratories

1980 -- The **Bayh-Dole Act** allows universities and other non-profit research organisations to automatically retain title to patents derived from federally funded R&D

1989 -- The **National Competitiveness Technology Transfer Act** allows contractor-operated federal labs (GOCO) to participate in CRADAs

1986 -- The **Technology Transfer Act** allows federal labs to co-operate in R&D with private firms (**CRADAs**) and to assign private firms the IPRs resulting from the cooperation

1996 -- The **Technology Transfer Improvements and Advancement Act** strengthens the rights of firms to exclusively licence patents resulting from CRADAs

1985/87 -- The National Science Foundation launches the **Engineering Research Centers** (ERCs) and the **Science and Technology Centers** (STCs) programmes to support industry-relevant research in universities

Other federal R&D at non-industrial performers

Federal R&D at universities

Federal R&D at other FFRDCs

Federal R&D at DOE FFRDCs

Licensing by three major Federal Agencies[1]

1996-98	Royalties recieved (million USD)	Number of licences granted	Of which non-exclusive	SME among licensees	Stock of active licences in 1998
DOE	0.44	31	28	23	73
NASA	1.47	111	39	94	108
NIH	102.2	607	514	346	990

DOE = Department of Energy NASA = National Aeronautics and Space Administration NIH = National Institutes of Health

Technology transfer indicators for universities and hospitals*

1998	Universities	Hospitals and research institutes
Number of new patent applications	4140	456
Number of patents issued	2681	407
Number of new licences	3078 (of which 1637 non-exclusive)	316 (of which 172 non-exclusive)
Number of spin-offs	279	26
Share of licence revenue[3]	28.69	60.47

* 132 universities and 26 hospital and research institutes that are members of the AUTM

Technology transfer indicators for Massachussets, the leading state for federally funded R&D[2]

	1991	1997
Number of new patent applications by non-business research institutions	227	651
Number of invention disclosures by non-business research institutions	635	1173
Invention disclosures by institution		
MIT		360
Massachusetts General Hospital		123
Harvard University		119
University of Massachussets		117
Brigham & Women's Hospital		86
Children's Hospital, Boston		68
Beth Israël -- NE Deaconess Hospital		65
Boston University		57
Dana-Farber Cancer Institute		54
Tufts University		40
Northwestern University		32
New England Medical Center		28
Brandeis University		25
Oceanographic Institute		3

Technology transfer indicators

1997	Share of license revenue[3]	Number of new licenses	Number of CRADAs[4]
All labs	1.62	242	1291
Brookhaven ◉	6.04	40	40
Oak Ridge ◉	5.51	38	238
Lawrence Livermore ○	3.76	19	237
Sandia ○	1.98	59	260
Stanford Accelerator ◉	1.69	3	2
Lawrence Berkeley ◉	1.52	56	140
Savannah River	1.22	5	17
Los Alamos ○	0.67	4	236
Argonne ◉	0.55	13	111
Ames ◉	0.24	5	10

○ Srong focus on defence-related R&D ◉ Strong focus on basic research

1. Concerns only government-owned inventions, not contractor-owned inventions (*e.g.* in 1997, DOE granted only 10 licenses, compared to 242 by DOE labs).
2. Massachusetts has the highest per capita federally funded R&D expenditures ($288) of the Leading Technology States (LTS), with the next closest LTS, California, at 64% of the Massachusetts level. Total federal R&D spending at Massachusetts non-profit research centers was $1.76 billion in 1997.
3. Per thousand dollars of R&D spending.
4. Cooperative Research Agreements with private firms; count for 1991, 1995 and 1997.

Source: OECD based on Jaffe (1999), GAO (1999), Massachusetts Technology Collaborative (1999), and AUTM (1999).

Increased patenting by universities and public labs has costs and raises *new issues*. Concurrent with this increase, the variety of ideas and research results that are being patented has also grown. This raises the risk of erosion in the social returns from public funding of research and of a possible decline in the quality of patents, and could have a negative impact on innovation in the private sector.

- The *expansion of patentable subject matter* (*e.g.* from life forms, DNA fragments, business methods to software, which hitherto relied on other forms of protection) *could in fact diminish the flows of ideas* and the diffusion of research knowledge in some disciplines.

- *Growing costs and risks of patent litigation are augmenting the uncertainty of innovation.* In addition, they incite industry to impose more stringent restrictions on publications of joint research. This contributes to lowering the quality of patents since potential innovators tend to file more patent applications in order to protect themselves from litigation. Increases in damages paid to plaintiffs create situations in which a patentee can gain more through litigation than through exploitation of its inventions. Excessive damages are a powerful deterrent to innovation, especially for small firms.

Globalisation

ISRs were structured around national research organisations and domestic firms at a time when the strategic interests of the different stakeholders converged easily towards national goals. Their international linkages were mainly through the scientific community with its longstanding tradition of global networking. The situation evolved gradually during the 1970s and 1980s, with the intensification of government-sponsored international co-operation in technological development, especially within Europe. The globalisation of firms' R&D strategies and access to public research (Box 4) together with the increased mobility of scarce highly qualified labour are now leading to more fundamental transformations:

- *The hierarchical and centralised model of ISR governance that prevails in a majority of countries must give way to a contractual and decentralised one.* Within public/private partnerships, the source of initiatives is shifting from government to firms; within governments, from central to regional and local authorities; within public research, from public labs to universities; and within public research organisations, from central management to labs and research teams. Since mission-oriented public research no longer plays a pivotal role within ISRs, new market-friendly co-ordination must be implemented, with greater involvement of the financial sector, especially venture capital.

- *Foreign firms often make more intensive use of public research than do domestic ones* (Box 5) and the efficiency of national support measures is enhanced when recipients are part of dynamic international networks. Governments must rethink how to maximise national benefits from ISRs that involve industrial participants taking a more global perspective. Building on globalisation to increase national benefits may require granting easier foreign access to national programmes and relaxing eligibility criteria regarding the location of publicly funded research activities, as well as greater international co-operation among governments to avoid opportunistic behaviour and distortion of competition.

- *Globalisation prompts publicly funded organisations to reconsider their role in the economy.* Some now enter into broad alliances with their homologues (Box 6) or private firms in order to create the knowledge platforms which look set to become key infrastructures of the knowledge economy.

Box 4. Globalisation of R&D and access to public research

Global innovation and technology management: the case of NESTEC

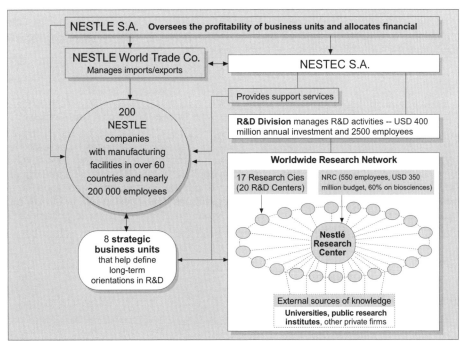

Source: OECD.

Globalisation of linkages between French firms and public research
Based on co-publications, 1998

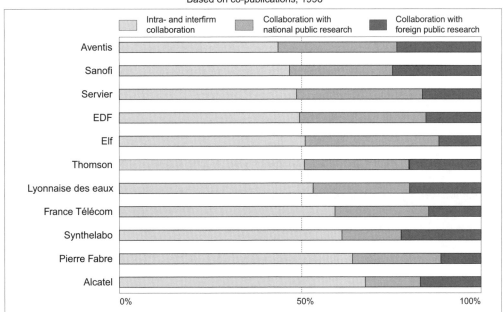

* The top eleven firms in terms of number of publications in 1998.
Source: Cance (2000).

Box 5. **A joint venture in Spain between public research and a foreign firm**

The National Centre for Biotechnology of the Spanish National Research Council (CSIC), was created to foster collaboration in biotechnology between industry and the public sector. The Centre attracted substantial funding from the Spanish subsidiary of Sweden's Pharmacia (now Pharmacia & Upjohn) in support of a joint seven-year research programme in cancer and immunology. A dedicated laboratory has been created at CSIC, staffed by personnel from both CSIC and Pharmacia & Upjohn. Intellectual property created in the laboratory belongs to the scientists concerned, but Pharmacia has priority commercial rights.

Box 6. **A government-sponsored partnership between MIT and Cambridge**

In mid-1998, the UK Chancellor of the Exchequer organised a series of seminars to assess the comparative performance of the UK economy. Among the lessons learnt was the key role that universities can play in improving UK performance. The idea for the Cambridge-MIT Institute (CMI) developed through subsequent discussions. The CMI will be active in four main areas: undergraduate student education; a programme of integrated research; adaptation of professional practice programmes developed at MIT to the UK experience; and creation of a national competitiveness network. The UK Government (DTI) has pledged to commit GBP 68 million over five years, while CMI plans to raise USD 25.5 million from the private sector.

NOTES

1. Gradual termination of privately run long-term programmes of research that can attain the size and financial stability necessary to create rewards and career paths for scientists and engineers paralleling those offered by universities or public labs (e.g. AT&Ts Bell Labs, IBM Cupertino's campus, Xerox's Palo Alto Research centre).

2. Sociologists, such as Callon in France, have gained a profound understanding of research networks and have developed techniques to quantify their characteristics, but this has mainly been achieved through case studies that are difficult to use in a systematic analysis of ISRs.

3. Columbia University is one of these exceptions, with a licensing income over one-fifth of its federal research funding.

BIBLIOGRAPHY

Association of University Technology Managers (AUTM) (1999), *AUTM Fiscal Year 1998 Survey*.

BMBF (Federal Ministry of Education, Science, Research and Technology) (1997), *Germany's Technological Performance*, December.

BMBF (Federal Ministry of Education, Science, Research and Technology) (2001), *Germany's Technological Performance 2000*, March.

Branscomb, L., F. Kodama and R. Florida (eds.) (1999), *Industrializing Knowledge: University-Industry Linkages in Japan and the United States*, The MIT Press, Cambridge, Mass.

Canadian Advisory Council on Science and Technology (1999), *Public Investments in University Research: Reaping the Benefits*, Report by the Expert Panel on Commercialisation of University Research, Ottawa, May.

Cho, M. (2000), "Safeguarding the Freedom of Research and the Broad Diffusion of Knowledge", paper presented at the German/OECD Conference on Benchmarking Industry-Science Relationships, Berlin, October.

Crance, M. (1999), "Publications : les onze ténors", *La Recherche*, December.

GAO (US General Accounting Office) (1999), "Technology Transfer: Reporting Requirements for Federally Sponsored Invention", August, http://www.gao.gov/audit.htm

Henderson, Rebecca, A. Jaffe and Manuel Trajtenberg (1998), "Universities as a Source of Commercial Technology: A Detailed Analysis of University Patenting, 1965-1988", *The Review of Economics and Statistics*.

Hicks, D. (2000), "Using Indicators to Assess Evolving Industry-Science Relationships", paper presented at the German/OECD Conference on Benchmarking Industry-Science Relationships, Berlin, October.

Howells, J., M. Nedeva and L. Georghiou (1998), *Industry-Academic Links in the UK*, PREST, University of Manchester.

Jaffe, A. and J. Lerner (1999), "Privatizing R&D: Patent Policy and the Commercialisation of National Laboratory Technologies", *NBER Working Papers Series*, No. 7064, Cambridge, Mass.

McMillan, G.S., F. Narin and D.L. Deeds (2000), "An Analysis of the Critical Role of Public Science in Innovation: The Case of Biotechnology", *Research Policy* 9, pp. 1-8.

Mansfield, E. (1998), "Academic Research and Industrial Innovation: An Update of Empirical Findings", *Research Policy* 26, pp. 773-776.

Massachusetts Technology Collaborative (1999), *Index of the Massachusetts Innovation Economy 1999.*

Mowery, D. (1998), "The Effects of Bayh-Dole on US University Research and Technology Transfer", paper presented at the OECD/TIP Workshop on Commercialisation of Government-funded Research, Canberra, 25 November.

Narin, F., K. Hamilton and D. Olivastro (1997), "The Linkages between US Technology and Public Science", *Research Policy* 26, pp. 317-330.

Narin, F., M. Albert, P. Kroll and D. Hicks (2000), *Inventing Our Future – The Links between Australian Patenting and Basic Science*, Australian Research Council, Canberra.

National Science Foundation (1998), *Science and Engineering Indicators 1998*, Arlington, VA.

NISTEP (National Institute of Science and Technology Policy) (2000), *Science and Technology (S&T) Indicators (2000 edition)* (in Japanese), NISTEP Report No. 66.

OECD (1999), *Managing National Innovation Systems*, OECD, Paris.

OECD (2000), *A New Economy? The Changing Role of Innovation and Information Technology in Growth*, OECD, Paris.

OECD (2001a), *The New Economy: Beyond the Hype*, OECD, Paris.

OECD (2001b), *Drivers of Growth: Information Technology, Innovation and Entrepreneurship*, special edition of the *OECD Science, Technology and Industry Outlook*, OECD, Paris.

Press, E. and J. Washburn (2000), "The Kept University", *The Atlantic Monthly*, March, pp. 39-54.

Rouach, D. and P. Santi (1999), "Case Study Nestlé: Interaction of R&D and Intelligence Management", in R. Boutellier *et al.* (eds.), *Managing Global Innovation*, Springer, pp. 274-286.

Schmoch, U., G. Licht and M. Reinhard (eds.) (2000), *Knowledge and Technology Transfer in Germany*, Fraunhofer IRB Verlag.

SPRU (2000), "Talent, Not Technology: Publicly Funded Research and Innovation in the United Kingdom", Science and Technology Policy Research (SPRU), University of Sussex.

Thorburn, L. (1999), "Institutional Structures and Arrangements at Public Sector Laboratories", paper presented at the TIP Workshop on High-technology Spin-offs from Public Sector Research, Paris, December.

US Council on Competitiveness (1998), *Going Global: The New Shape of American Innovation*, September.

Chapter 2

BENCHMARKING INDUSTRY-SCIENCE RELATIONSHIPS

National systems of ISRs

Globalisation and the diffusion of best practice policies have mitigated some of the differences in national systems of industry-science relationships (ISRs), thus changing their comparative advantages, but they have not eliminated the considerable diversity of the models implemented. The interactions between the public research sector and industry in different countries take various institutional forms and differ in their nature and intensity, reflecting national specificities in institutional set-ups, regulatory frameworks, research financing, intellectual property rights and in the status and mobility of researchers. Existing internationally comparable indicators capture some of these differences. Measurable national differences with implications for industry-science linkages include: *i)* the institutions responsible for performing and funding research development; *ii)* the trends driving the funding and performance patterns of R&D; and *iii)* specialisation in specific scientific disciplines.

Figures 1 and 2b show an extremely wide dispersion in rates of government funding (from more than two-thirds in Mexico to less than one-fifth in Japan) as well as in shares of publicly funded organisations in R&D performance (from over two-thirds in Greece, Mexico and New Zealand to less than one-quarter in Ireland, Japan, Korea, Sweden and the United States). The roles of the two main types of publicly funded organisations (universities and research institutes) in R&D performance vary even more, although the share of universities has been steadily increasing in most countries in the last decade. Figure 3 highlights wide disparities in terms of the principal actors of business R&D and the orientation and financing of the public research sector. The OECD countries fall into four categories and ten sub-categories:

- Countries with a very high share of government R&D funding and performance:

    - University-based system (Turkey)
    - Broad-based system (Italy, New Zealand, Poland, Portugal, Mexico)
    - Institute-based system (Hungary, Iceland)

- Countries with a moderately high share of government R&D funding and performance:

    - University-based system (Austria, Spain)
    - Broad-based system (France, Netherlands, Norway)

- Countries with an average share of government R&D funding and performance:

    - University-based system (Canada, United Kingdom)
    - Broad-based system (Denmark, Finland, Norway, Germany)
    - Institute-based system (Czech Republic)

- Countries with a low share of government funding and performance:

    - University-based system (Belgium, Ireland, Japan, Sweden, Switzerland, United States)
    - Institute-based system (Korea)

Figure 1. Share of publicly funded organisations* (PFOs) in R&D performance
1998, percentage

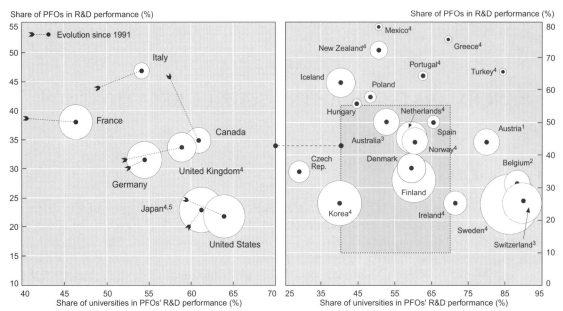

* Non-business R&D performers, excluding non-profit private organisations.
Note: Circles are proportionate to countries' relative R&D intensity (total R&D expenditures as % of GDP), with a maximum for Sweden (3.8%) and a minimum for Mexico (0.3%).
1. 1993.   2. 1995.   3. 1996.   4. 1997.   5. Underestimated.
Source: OECD.

Figure 2a. Share of business in the funding of research performed by government and university
1998 or latest year available, percentage

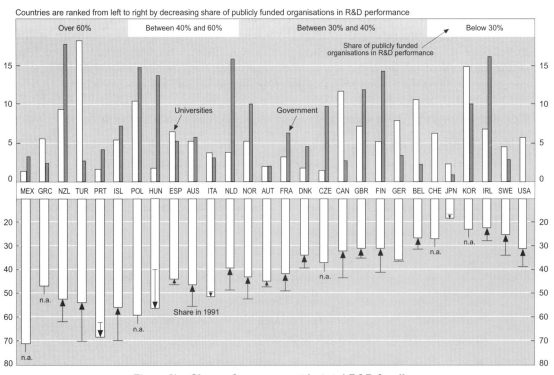

Figure 2b. Share of government in total R&D funding
1998 or latest year available; percentage

Source: OECD.

Figure 3. **Knowledge production structures in selected countries at the end of the 1990s**

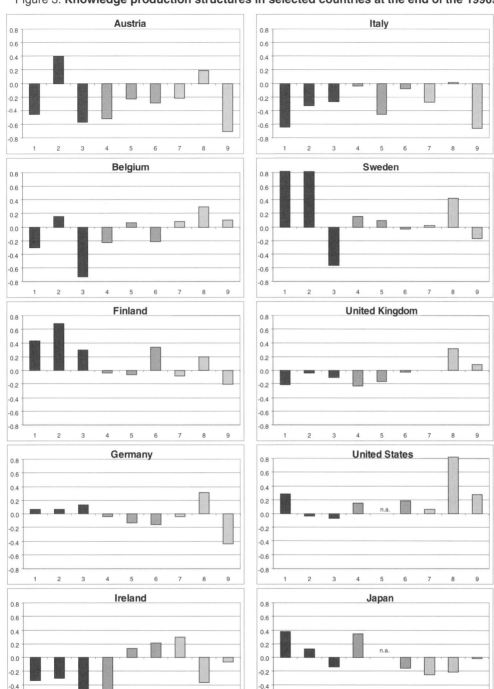

Note: All values represent the deviation from the EU average.

R&D orientation	**Enterprise sector structure**	**Public research sector structure**
1. BERD in % of GDP	4. Very large enterprises in BERD (%)	7. Natural sciences and engineering (NSE) in HERD (%)
2. HERD in % of GDP	5. Continuously R&D-performing SMEs (%)	8. Impact factor of scientific publications
3. GOVERD in % of GDP	6. BERD performed in high-tech sectors (%)	9. HERD financed outside general university funds (%)

Source: Austrian Federal Ministry of Economy and Labour (2001), based on OECD and other sources.

In broad terms, the main challenge for the first group of countries, those with below-average R&D intensity, is to increase the technological absorption capacity of firms and thus shift a greater proportion of R&D activity to the private sector. Countries in the second and third groups must seek to improve ISRs with the overriding goals of reducing unnecessary duplication of innovation investment and improving the responsiveness of the public sector to the needs of industry. In the last category, the overriding concern is to cultivate excellency in university research and increase the leverage of the relatively low level of public investment in research.

The challenges faced by countries and the "implementability" of different types of responses will also vary depending on some of the more subtle features of national research systems. The United States, the United Kingdom, France and Sweden need to maximise economy-wide spillovers of declining – although still sizeable – defence-related R&D investments. There are also important differences across countries regarding not only the size but also the content of research activities in universities and public institutes. In English-speaking and Scandinavian countries – but also in Japan and Portugal – universities conduct the bulk of basic research, while the public research institutes focus more on applied research missions. In continental Europe, university research co-exists with R&D by public sector laboratories, with both types of institutions performing basic research and mission-oriented activities, thus increasing the danger of duplicating research efforts.

National science systems support innovation by generating new, economically relevant knowledge and by facilitating absorption of knowledge generated in foreign countries – the balance between these two functions varies with country size and S&T specialisation. Scientific specialisation profiles differ substantially across countries, are more contrasted in small than large countries and tend to be fairly stable over time (Figure 4). Although their transformation might be one of the desirable long-term outcomes of improved ISRs, they must be taken almost as a given when considering options to trigger such improvement.

In small and medium-sized countries, scientific output in industry-relevant disciplines is well correlated with R&D intensity, with only a few outliers such as Korea, where R&D performance is disconnected from scientific output (Figure 5a). Larger countries seem to enjoy economies of scale in translating scientific efforts into R&D, except the United Kingdom, where scientific output is inflated by prolific publishing by the medical sector, and Italy. In fact, under-specialisation in science-intensive industries explains in large part why R&D intensity is over proportionate to scientific output in Japan and Germany. This is confirmed by Figure 5b, which shows that the "link" (measured as the number of patents) between science and patentable innovation is weaker in these countries that in other G7 countries, except Italy. Figure 5c suggests that, in Japan more than in Germany, an additional explanation is the relatively low productivity of the science system, as measured by citations of scientific papers.

Figure 4. **National profiles of relative scientific specialisation**

Based on publications; 1998

ⓧ Biosciences, medical, clinical and pharmaceutical research

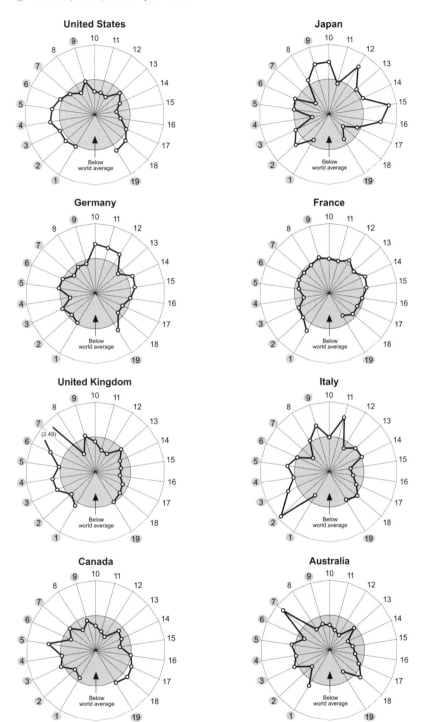

1. Microbiology & virology
2. Oncology
3. Gastroenterology & cardiology
4. Epidemiology, public health
5. Neurosciences, neuropathology
6. Medicine, miscellaneous
7. General & internal medicine
8. Analytical chemistry
9. Medical chemistry & pharmacy
10. Chemistry
11. General & nuclear physics
12. Applied physics
13. Optics, electronics, signal processing
14. Physical chemistry, spectroscopy
15. Materials science, metallurgy, crystallography
16. Chemical engineering, polymer science
17. Mechanical engineering, fluid dynamics
18. Computer & information science
19. Biomedical engineering

Source: OECD, based on data from OST.

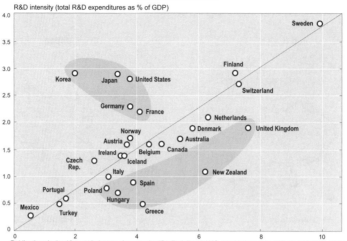

Figure 5a. **R&D intensity and scientific output in industry-relevant fields***

1998 or latest year available

* For a list of the 19 scientific disciplines, and relevant specialisation profiles of G7 countries, see Figure 6.
Source: OECD, partly based on data from OST.

Figure 5b. **Science linkage* and scientific output**

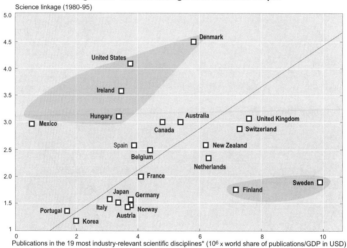

* Measured by the average number of scientific publications in industrial patents.
Source: OECD, based on data from CHI-Research and OST.

Figure 5c. **Productivity of the science system* and scientific output**

1998 or latest year available

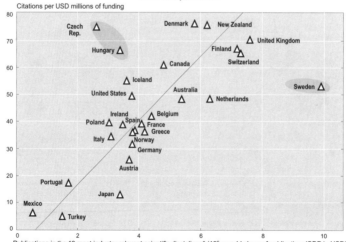

* Measured by citations per USD million of funding.
Source: OECD, based on data from OST and SPRU (2000).

Conceptual framework and indicators

Traditionally, the relationships between publicly funded research and industry have been analysed based on simplistic models, thus directing attention away from the issues that have grown in importance over time to become the most crucial today (SPRU, 2000).

A frequent simplification is to equate universities to public research, public research to science, innovation to proprietary technologies, and assume a linear relationship between science and innovation. This linear model cannot really explain why some innovation systems perform better than others and *a fortiori* provides no guidance for a comparative evaluation of ISRs. Modern innovation theory sees innovation as a process rather than a product, and stresses the complex feedback systems between basic research and industrial R&D. It also recognises that publicly funded research institutions are diverse, comprising different types of universities and public labs, whose missions may overlap since they are products of an historical, evolutionary process and not the results of rational exercises by welfare-maximising public authorities (David and Foray, 1995). ISRs are not simply transactions mirroring a clear-cut division of labour in the production of knowledge. Rather, they represent an institutionalised form of learning that provides a specific contribution to the stock of economically useful knowledge. They should be evaluated not only as knowledge transfer mechanisms but also on their other capacities (*e.g.* building networks of innovative agents, increasing the scope of multidisciplinary experiments). To this end, industry-science linkages must be characterised along three dimensions: nature and relative importance of the channels of interaction; their institutional arrangements; and their incentive structures, as influenced by government's promotion programmes (Table 1 presents the results of an Austrian/European Commission benchmarking of these programmes).

Figure 6. **A conceptual framework for assessing industry-science relationships**

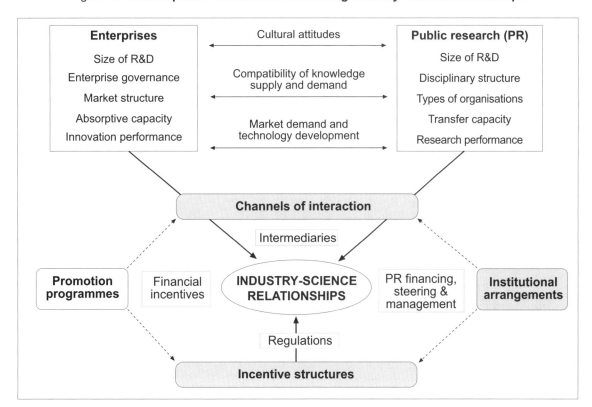

Source: OECD, adapted from Polt *et al.* (2001).

Table 1. ISR-related promotion programmes in selected countries at the end of 1990s

		Austria	Belgium	Finland	Germany	Ireland	Italy	Sweden	UK	US	Japan
Overall financial support	Share of government funding in BERD in %	9.8	4.4	6.3	8.0	5.3	13.3	7.6	11.6	14.4	1.3
	ISR promotion programmes in ‰ of GDP (estimates)	~0.3	~0.2	~0.9	~0.4	~0.3	~0.1	~0.4	~0.3	~0.5	n.a.
	ISR promotion progs. in % of government R&D financing	~5	~5	~11	~4	~11	~2	~4	~5	~5	n.a.
Public promotion programmes	R&D financing for enterprises ("indirect ISR promotion")	=	=	=	=	>	>	=	=	=	=
	Tax allowances to enterprises for ISR activities	=	=	=	=	=	>	=	=	=	=
	Specific financial support for collaborative/contract research	=	>	>	>	>	=	>	>	=	=
	Specific support to SMEs for ISR activities	=	>	>	>	>	=	>	>	=	=
	Support for joint R&D facilities	>	>	>	=	>	=	>	>	=	>
	Technology focus (centres of expertise, etc.)	=	=	>	>	>	>	>	>	=	>
	Support for researcher mobility	=	=	=	=	=	=	=	=	=	=
	Support for (under)graduate training at enterprises	=	=	>	=	=	=	=	>	=	=
	Promotion of employees training in HEIs	=	=	=	=	=	=	=	>	=	=
	Promotion of co-operation in curricula/education planning	=	>	>	=	=	=	=	=	=	=
	Raising transfer capacities in public science institutions	>	>	>	>	>	>	>	>	=	>
	Support to public science researchers for IPR activities	>	=	>	>	=	=	=	>	=	=
	Start-up promotion in HEIs/PSREs	>	>	>	>	>	>	>	>	=	=
	Promotion of networking initiatives	>	=	>	>	>	=	>	>	=	=
	Awareness measures both in industry and science system	=	=	=	>	=	=	>	>	>	=
	Regional approaches to ISR promotion	=	>	>	>	=	>	>	>	=	=

■ : high relevance; ■ : important; □ : less important/missing; > : currently gaining public attention; = : no major current change in public attention.

Source: Austrian Federal Ministry of Economy and Labour (2001).

ISR channels include contract research, consultancy and services, IP transactions, knowledge spillovers,[1] co-operation with firms for teaching/training, labour mobility. The institutional arrangements for ISRs can be considered from a macro perspective (type and respective role of publicly funded research organisations, degree of intermediation) or a micro perspective (the legal and organisational framework for the management of the linkages by individual research or intermediary organisations). The incentive structures are of a financial or regulatory nature and can be analysed at both the macro and micro levels, depending on whether they are established by government or by the management of individual organisations. Benchmarking requires indicators, but the availability of quantitative indicators on ISRs is still quite poor (Tables 2 and 3), and existing statistics must be used with great care since similar concepts may mean different things in different national institutional settings.

Table 2. **Availability of benchmarking indicators**

		Mainly qualitative	Quantitative		
			Fair	Mostly country-specific	Poor in most countries
Structure of ISRs	*Size and orientation of public research (PR)*		♦	♦	
	Size and functions of intermediaries	♦			
	Absorptive capacity of the business sector			♦	♦
	Government incentives and programmes	♦		♦	
	Regulatory framework	♦			
Intensity of ISRs	*Financial flows between public and private research organisations*				
	▪ Overall flows			♦	
	▪ Business R&D contracted out to PR			♦	♦
	▪ Contract-based financing of PR			♦	♦
	Labour mobility				♦
	Other knowledge flows (*e.g.* joint publications, patents, co-operative R&D)		♦	♦	
Economic impact of ISRs	*Macro-indicators*	♦			
	Firm-level indicators		♦	♦	
	Structural indicators				
	▪ Spin-offs and other NTBFs				♦
	▪ Research-based networks and clusters				♦

Source: OECD.

Table 3. Indicators of ISRs in selected countries at the end of 1990s
Shaded fields: values significantly above EU average

Type of ISR	Indicator	Austria	Belgium	Finland	Germany	Ireland	Italy	Sweden	UK	US	Japan
Contract and collaborative research (1998)	R&D financing by industry for HEIs in % of HERD	2.0	10.6	4.2	9.7	6.4	3.8	4.5	7.2	6.0	2.4
	R&D financing by industry for PROs in % of GOVERD	2.0	2.1	14.0	2.0	15.4	3.0	2.9	11.9	n.a.	0.9
	R&D financing by industry for HEIs/PROs in % of BERD	1.7	4.9	3.9	2.9	3.4	3.2	1.5	5.0	1.7	0.6
Faculty consulting with industry	Significance of R&D consulting with firms by HEI researchers	high	low	low	high	low	low	n.a.	high	high	high
	Significance of R&D consulting with firms by PRO researchers	low	low	low	low	low	low	n.a.	low	high	high
Co-operation in innovation projects (1996)	Innovative manuf. enterprises that co-operate with HEIs (%)	12.6	13.4	47.3	10.4	13.8	2.5	26.1	11.3	n.a.	n.a.
	Innovative manuf. enterprises that co-operate with PROs (%)	7.1	8.5	38.0	13.6	6.3	1.3	16.3	4.5	n.a.	n.a.
	Innovative service enterprises that co-operate with HEIs (%)	5.8	15.3	19.2	7.2	3.6	n.a.	12.0	2.9	n.a.	n.a.
	Innovative service enterprises that co-operate with PROs (%)	2.5	6.0	13.8	3.0	2.5	n.a.	5.8	21.9	n.a.	n.a.
Science as an information source for industrial innovation (1996)	Innov. man. ent. that use HEIs as inform. source in innov. (%)	4.7	6.7	6.9	6.7	5.0	1.7	4.5	3.9	n.a.	n.a.
	Innov. man. ent. that use PROs as inf. source in innov. (%)	1.1	4.8	5.3	2.9	7.4	1.6	n.a.	1.9	n.a.	n.a.
	Innov. serv. ent. that use HEIs as inform. source in innov. (%)	0.6	2.0	2.7	5.6	5.8	n.a.	4.7	3.7	n.a.	n.a.
	Innov. serv. ent. that use PROs as inf. source in innov. (%)	0.7	2.7	0.6	2.7	2.1	n.a.	n.a.	6.9	n.a.	n.a.
Mobility of researchers	Researchers in HEIs moving to industry p.a. in %	medium	~3	~3.5	~5	low	low	~4	high	>2	low
	Researchers at PROs moving to industry p.a. in %	medium	~5	~4	~3	low	low	~15	medium	medium	low
	HE graduates in industry moving to HEIs/PROs p.a. in %	low	0.4	0.4	medium	low	low	0.6	low	medium	high
Training and education	Income from vocational training in HEIs in % of R&D exp.	low	high	9	low	medium	low	n.a.	2.5	high	low
	Vocational training particip. in HEIs per R&D empl. in HEIs	low	high	16	low	medium	low	n.a.	high	high	high
	Share of students employed in enterprises during their studies (placements, masters' thesis, PhD programmes) in %	medium	high	high	high	low	medium	n.a.	high	high	high
Patent applications by public science	Patent applications by HEIs (and individual HEI researchers) per 1 000 employees in NSEM* in HEIs	low	high	high	19	low	low	n.a.	~15	>35	5 - 10
	Patent applications by PROs (and individual PRO researchers) per 1 000 employees in NSEM* at PROs	medium	~15	~12	20	low	low	n.a.	medium	>15	low
Royalty incomes by public science	Royalties in % of total R&D expenditures in HEIs	low	low	~0.3	~0.7	low	low	n.a.	~0.5	2.3	<0.01
	Royalties in % of total R&D expenditures at PROs	low	low	low	low	low	low	n.a.	medium	0.15	low
Spin-offs from public science	Technology-based start-ups in HEIs per 1 000 R&D personnel	~4	<1	2 - 3	3 - 4	low	low	n.a.	high	>3	low
	Technology-based start-ups at PROs per 1 000 R&D personnel	~1	~3	~1	2 - 3	low	medium	n.a.	medium	medium	low
Informal contacts, personal networks	Significance of networks between industry and HEIs	medium	low	high	high	low	low	high	high	high	high
	Significance of networks between industry and PROs	high	high	high	medium	low	low	n.a.	high	high	high

* Natural sciences, engineering (including agricultural sciences) and medicine.
Figures refer to the latest year available, which is normally 1997, 1998 or 1999. In the case of missing data, assessments by national experts are given.
HEIS = Higher Education Institutions; HERD = R&D expenditures by HEIs; GOVERD = Government R&D expenditures; PROs = Public laboratories; BERD = Business R&D expenditures.
Source: Austrian Federal Ministry of Economy and Labour (2001), based on data from the OECD, the EU, and various national sources.

Benchmarking selected channels, incentive structures and institutional arrangements of ISRs

Channels of ISRs: the examples of spin-offs and labour mobility

Spin-offs from public research

Spin-offs are: *i)* firms founded by public sector researchers, including staff, professors and postdocs; *ii)* start-ups which have licensed public sector technologies; and *iii)* firms in which a public institution has an equity investment, or which were directly established by a public research institution. Spinning off is the entrepreneurial route to commercialising knowledge developed by public research and as such is attracting a great deal of attention, given the current "start-up fever" in many countries. However, governments, too, have a special interest in this specific type of industry-science linkage because it may be one of the factors that explain differences in performance in new, fast-growing science-based industries, especially biotechnology. In addition, some are tempted to see the spin-off formation rate as a key indicator of the quality of ISRs, prompting public research organisations to place greater priority on this aspect of their commercialisation strategy and to publicise their achievements in this area. However, this growing policy interest has revealed the paucity of the information that is needed to judge whether spin-offs really warrant such attention, to which extent and why their rates of formation are increasing and differ across countries, and how best governments should promote them. Results from the OECD benchmarking project suggest the following answers:

- The main contribution to innovation of spin-offs from publicly funded research is not direct and is more qualitative than quantitative. The number of such firms born each year remains very modest compared to corporate spin-offs (*i.e.* a few hundred compared to several thousand), which themselves represent only between 10% and 30% of total technology-based start-up activity in European countries. As a channel of ISRs, their role should also be put in perspective. In the United States, they accounted for just over 10% of the technology licences negotiated by universities in 1998 – which is a modest share, but far larger than their relative weight in the total of new technology-based start-ups. This confirms other indications that they probably play a different role than other new technology-based firms in the innovation system, as vital components of clusters of innovative firms formed around academia and industry, and of social networks in science-based industries (Mustar, 2000).

- *The number of spin-offs generated per public institution or per country is generally on the rise*, although some countries seem to have already experienced a peak in spin-off formation in the late 1980s or early 1990s. In the small sample of countries for which data are available, France stands out as an exception in that it exhibits declining public sector entrepreneurship during the 1990s. In some countries, spin-off formation seems to have a cyclical dimension (Figure 7).

- *The public research sector generates more spin-offs in some countries than in others.* International benchmarking of the rates of spin-off formation is a difficult task due to the lack of comparable data. However, provisional crude estimates by the OECD Secretariat for a small sample of countries demonstrate that in the 1990s this rate was about three to four times higher in North America than in most other OECD countries, although current dynamic developments are rapidly changing this picture (Table 4).

Table 4. **Spin-off formation in the 1990s in selected OECD**

		Australia[1]	Belgium[2]	Canada[3]	Finland[4]	France[5]	Germany[6]	United Kingdom[7]	United States[8]
	Coverage	All publicly funded organisations	All publicly funded organisations	Universities	Public lab (VTT)	All publicly funded organisations	Public labs	Universities	Universities
Cumulative	Period	1971-99	1979-99	1962-99	1985-99	1984-98	1990-97	1984-98	1980-98
	Number	138	66	746	66	387	462	171	1 995
Per year	Period	1991-99	1990-99	1990-98	1990-99	1992-98	1990-97	1990-97	1994-98
	Number	10	4	47	5	14	58	15	281
	per 10^9 USD of R&D	3.3	3.6	7.4	n.a.	2.5	n.a.	n.a.	12

1. Narrow definition (public employee among founders and licensed technology from public sector).
2. Very broad definition (any firm created to commercialise a research result of a university or technical school).
3. Broad definition (includes any firm created as a result of the externalisation of a service of a university department).
4. Broad definition.
5. Broad definition (any firm founded by a university or public lab employee professor, post-doc or research alumni).
6. Broad definition.
7. Very narrow definition (includes only firm in which university makes an equity investment).
8. Narrow definition (includes only firms dependent on the licensing of technology from a university for their initiation).

n.a.: not applicable because of a too broad or too narrow definition of spin-offs.

Source: OECD.

Table 5. **Distribution of spin-offs by sector or area of scientific expertise**

Australia		Canada		Finland (VTT)		France (CNRS)		United Kingdom	
Biotech	35%	Biotech & pharma	24%	Electronics	21%	Informatics	25%	Engineering	20%
ITT	22%	Medical	18%	Manufacturing	21%	Health	20%	Biotech	19%
Non high-tech	21%	Software	16%	IT	15%	Instrumentation	8%	Software	11%
Instrumentation	9%	Electronics	11%	Automation	15%	New materials	7%	Chem./physical	11%
New materials	5%	Communications	5%	Energy	12%	Electronics	7%	Consultancy	10%
Pharmaceuticals	4%	Agri-food	4%	Building	8%	Environment	6%	Life sciences	9%
Aerospace	1%	Chemicals	3%	Biotech & food	3%	Accoustics/optics	5%	Medicine	5%
Unknown	3%	Others	19%	Others	5%	Others	22%	Others	19%

Source: OECD, from various sources.

Figure 7. **Stylised trends in spin-off formation**

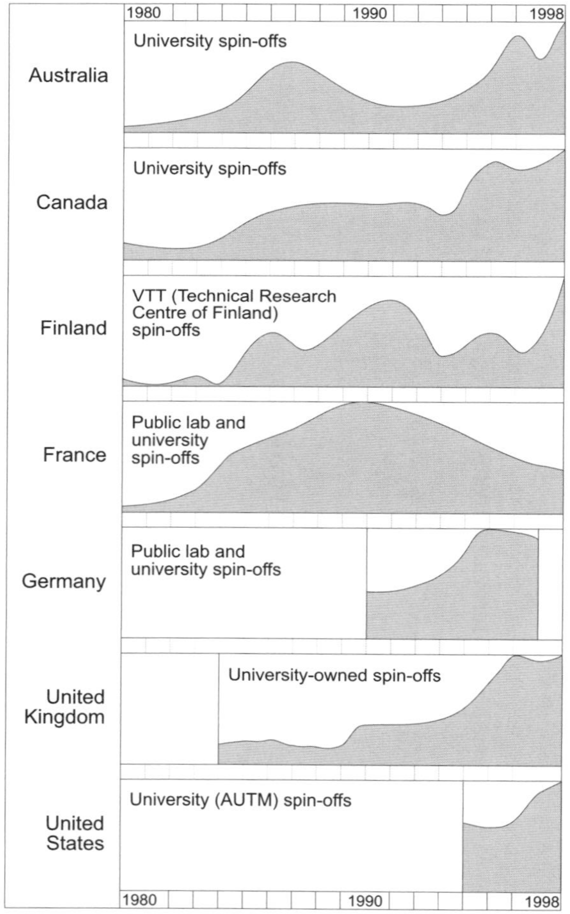

Australia — University spin-offs
Canada — University spin-offs
Finland — VTT (Technical Research Centre of Finland) spin-offs
France — Public lab and university spin-offs
Germany — Public lab and university spin-offs
United Kingdom — University-owned spin-offs
United States — University (AUTM) spin-offs

Source: OECD, from various sources.

- *General conditions for entrepreneurship cannot explain all the international differences in the rates of spin-off formation.* These differences almost mirror those in overall start-up rates when comparing the United States, Canada and France (respectively 8.4%, 6.8% and under 2% of the firm population according to GEM).[2] However, countries such as Finland and Germany, both of which have strongly diffusion-oriented innovation systems, spur more public spin-offs than one would expect based on GEM indicators.

- Spin-offs from public research are generally heavily concentrated in the information technology and, increasingly, the biotechnology/medical technologies sectors (Table 5). They are as much an indicator of public sector activity in these areas as of its entrepreneurship.

- *Setting spin-off performance targets is ill advised.* Even with data that are normalised by researchers or research budgets, cross-country comparisons must be performed with great caution. The purpose of a benchmarking exercise should not be to try to develop spin-off "targets" for countries or institutions. First, the types of research institutions that make up the national research base in each country are too varied. Second, the importance of public spin-offs to an economy and as a performance indicator for a public research organisation must be assessed in the context of other technology transfer mechanisms – sale and licensing of technology, contract or collaborative research, and human mobility.

All governments are aware that improving the environment for entrepreneurship will help to foster the generation of public research-based spin-offs. The real issue is whether more targeted promotional policies are warranted. On the one hand, policy makers need to decide how much they want to invest in a mechanism that favours particular industries rather than new-firm creation as a whole. On the other, the vitality of the public research sector is at stake, and success in these industries cannot wait for changes in the entrepreneurial climate, especially as these may take a long time. In addition, the experience of some countries suggests that there are specific obstacles to public research-based spin-offs that only government can lower. Public seed capital to help finance early-stage investment when uncertainty is too high and the size of projects too small for private venture capital, has proved useful, especially in countries where informal investors ("business angels") cannot contribute much to filling the gap. Nevertheless, the main role of government is to improve institutional frameworks (*e.g.* incubators, management of public research organisations) and incentive structures (*e.g.* regulations governing researchers' mobility and entrepreneurship).

Table 6. **Mobility of employees with higher education in three Nordic countries** (1994-95[1])

Delivering sector (% share)							% of all employees with higher education		Rate of internal mobility		Receiving sector (% share)				
Outside workforce	Public services	Private services	Industry	HE or RDI	Rate					Rate	HE or RDI	Industry	Private services	Public services	Out of workforce
Mobility in							**Higher education (HE)**				*Mobility out*				
64.5	24.6	6.7	2.0	2.2	27.3%		6.1%	Finland	11.4%	21.5%	3.0	11.3	9.3	23.1	53.5
43.9	38.1	10.5	2.7	4.8	13.3%		4.4%	Norway	4.2%	14.4%	8.6	7.0	12.7	24.4	47.3
8.4	64.0	16.4	7.9	3.3	14.5%		5.9%	Sweden	3.7%	18.2%	26.1	9.9	12.6	28.1	23.3
Mobility in							**R&D institutes (RDI)**				*Mobility out*				
53.5	17.5	9.2	6.0	13.8	13.3%		1.8%	Finland	7.9%	12.3%	15.0	19.6	12.9	12.0	40.5
36.9	18.9	13.8	5.7	24.7	12.0%		2.1%	Norway	2.7%	16.4%	10.4	11.4	29.6	14.6	34.0
2.5	6.0	9.5	10.2	71.8	32.2%		0.8%	Sweden	2.8%	19.8%	10.3	34.9	32.8	9.9	12.1

1. 1995-96 for Finland.
Source: OECD, based on data from the NIS Focus Group on Mobility; see http://www.oecd.org/sti/innovation

Labour mobility

Low rates of mobility of scientists and researchers remain major obstacles to improving industry-science linkages in a number of OECD countries. In some, public researchers find themselves in a "public employment trap" whereby low industry funding of R&D (and thus weak private demand for researchers), combined with regulatory barriers and disincentives to mobility, result in a concentration of researchers in the public sector. The risk of such a trap is less in countries where industry funds and performs a greater share of R&D and where wage competition for skills spurs movement from the public sector to industry.

General data on job mobility, based on average job tenure, suggest that overall mobility is higher in Australia, the United States, the United Kingdom, Canada and the Netherlands, and lower in Italy, Belgium, France, Finland and Japan. International statistics on the mobility of researchers and scientists are too scarce to allow international comparisons. However, country data on job changes among scientists and researchers do provide some indication of the role of mobility in innovation systems:

- In the United States, for example, scientists and engineers change jobs every four years, and even more frequently in the case software and IT occupations. In Japan, it is estimated that only 20% of engineers change jobs once in their career and it is likely that job changes between the public and private sectors are even less frequent, given the tradition of lifelong employment in industry and the existence of restrictive regulations on university professors' interactions with industry. In the United Kingdom, public sector researchers with short-term contracts play a key role in the transfer of knowledge from the public to the private sector.

- The most direct evidence on the mobility of qualified labour is provided by register data in the Nordic countries. Table 6 shows that flows from universities to industry are quite low but are higher in Finland and Sweden than in Norway. For all three countries, the public services sector receives most of the annual outflows of science and engineering personnel from the higher education sector, suggesting strong demand by hospitals. With regard to movements of science and engineering personnel out of R&D institutes, industry and services are the main destination in all three countries although, in the case of Norway, movement to services is greatest.

- The scope for fostering mobility also differs between research institutions and universities and tends to be higher in the latter. Mobility of university researchers is relatively high in Germany where, each year 5-6% of all university researchers move to industry and 3-4% move to public laboratories, above the average in the European Union (Rammer, 2001). One of the main drivers of this mobility in Germany, but also in the United Kingdom, is the temporary employment conditions offered to younger researchers. In 1997, Japan introduced a fixed-term system in employment at universities and the national testing and research institutes to foster greater mobility. In France, each year only about 40 scientists leave their public research organisations to work in industry. In fact, temporary flows between industry and public research organisations are more common than permanent moves. Mobility is also lower or higher depending on scientific discipline; data on Dutch mobility show that researcher mobility is lower in the humanities and the social sciences.

Employment regulations and conditions in the labour market set the overall pre-conditions for changing jobs and occupations. Flexibility in labour markets (including wages) can facilitate occupational and geographic mobility. In addition, the lack of transferability of pension schemes between the public and private sectors is a major barrier to the mobility of researchers in many OECD countries. More specific regulatory constraints include:

- *Public employment legislation* insofar as a large share of researchers in Italy, Spain, Portugal but also in Australia, Denmark and, to a lesser extent, France, work in the higher education and government sectors. Until recently for example, public researchers in Japan and France were explicitly prohibited from undertaking activities with industry due to their civil servant status. This was also the case for Italian public sector researchers until 1999, when a new law was passed that allows public researchers to work in firms on a temporary basis. In some countries, residency requirements – the requirement to live in a certain region in order to be eligible for public employment or professional practice – also limit mobility.

- *Regulations governing temporary mobility* (*i.e.* secondments and staff exchanges) that are generally institution-specific. In most European countries, university secondment and sabbatical provisions mainly concern research in other public research institutions, although increasingly universities such as those in the United Kingdom emulate initiatives in Canadian and US universities which allow professors to take leave to work in industry. Even in these countries, however, temporary movements of research personnel tend to be one-way, from university to industry.

- *Regulations regarding remunerative secondary employment* for public researchers also tend to be institution-specific, except in the case of national university systems or where researchers are public employees. German university professors and public sector researchers, for example, are allowed remunerative secondary occupations (normally limited to 20% of working time), sometimes dependent on administrative approval. In Finland, the researchers at the Academy of Science must request approval for temporary outside appointments or take leave to run a business or conduct co-operative research with industry.

- *Regulations affecting academic entrepreneurship* mainly seek to limit the amount of time a researcher is involved in the day-to-day activities of the firm and the potential for conflict between the research institution's interests and the monetary interests of the researcher. In some countries, public researchers, whether civil servants or not, are prohibited from serving on the board of a private company. In Belgium, professors are allowed to sit on the board of directors and be company stakeholders, but are forbidden to actively hold a director's position in the company or to receive remuneration from their industrial activities. In Hungary, public sector researchers must disclose their entrepreneurial activities and permission is granted on a case-by-case basis.

While such limitations may be grounded in sound considerations such as avoiding conflict of interest and ensuring that teaching missions are appropriately carried out, it is the way in which regulations are implemented in practice that has a greater impact on outcomes. In many cases, the possibility for leave, especially for longer periods, depends on finding a suitable replacement. In addition, temporary leave tends to be reserved for tenured professors or public researchers with permanent employment, although the propensity for mobility generally decreases with age across all occupational categories. Finally, simply allowing researchers the option to hold secondary employment may not be a sufficient incentive if it is not accompanied by changes to the way in which promotion and rewards are granted. Often, with the notable exception of polytechnics, vocational and engineering schools, university reward systems do not sufficiently recognise non-academic activities.

Despite the persistence of barriers such as the portability of pensions, there is a clear trend across OECD countries towards relaxing regulatory constraints on mobility and academic entrepreneurship. On the one hand, governments such as in Austria and Finland are granting more autonomy to universities; on the other, they are relaxing rules on public research collaboration with industry. Under the new innovation law in France, the temporary engagement of public sector researchers and secondary remuneration is permitted. Italy instituted new laws in 1999 that allow greater mobility of researchers to the private sector, especially to SMEs, via a system of temporary appointments. In 2000, the UK Government published draft guidelines on changes to the Civil Service Management Code (concerning the employment contracts of civil servants), allowing civil service employees to benefit from incentive-based pay linked to the commercialisation of research results. Canada is implementing the 1995 internal trade agreement that abolishes residency requirements for employment in the provinces and calls for the mutual recognition of occupational qualifications across provinces.

Box 1. Promotion schemes for researcher co-operation with industry

Austria maintains mobility promotion schemes such as "Scientists for the Economy" and the mobility of junior researchers is promoted through the Industrial Promotion Fund.

Australia's Strategic Partnerships with Industry-Research and Training Scheme and the Co-operative Research Centres Programme are aimed at improving public-private mobility and co-operation.

Canada: The National Science and Engineering Research Council sponsors postgraduate training in industry through various schemes including scholarships for training masters and doctoral students in industry and fellowships for the hiring of a recent PhD graduate by a firm.

France: The Ministry of Research fosters PhD training in a research company by subsidising up to half of the corresponding salary costs to the firm. Subsidies for post-doctoral positions in SMEs are available to young PhDs without industry experience.

Japan's latest Basic S&T Promotion Plan outlines a series of regulatory reforms to the labour market for public, sector research, aimed at improving mobility between the public and private research sectors. The Centres for Co-operative Research in 56 national universities carry out joint industry-public research as well as technical training of researchers from private companies. A main goal is to create critical mass by canalising individual researcher collaboration into institutional level linkages.

Korea: The Korean Institute of Science and Technology (KIST) has promotional schemes to grant temporary leave to researchers to undertake entrepreneurial activities.

The *Netherlands'* KIM scheme that promotes the movement of S&T personnel to SMEs has proved to be successful. Furthermore, under the WBSO (Act to promote R&D), small firms are allowed a tax deduction for the labour costs of R&D staff.

Norway has set up special programmes (such as the FORNY programme that is now entering its third phase) to stimulate mobility from universities/research institutes to the private sector and to make industry-relevant research more attractive.

Portugal: The Ministry of Science and Technology runs a programme to help the placement of new PhDs in firms through the subsidisation of salaries for two years.

Sweden: The NUTEK competence centres at universities promote collaboration between public researchers and those in firms which may help break down non-regulatory barriers to mobility.

United Kingdom: The Faraday Programme promotes a continuous flow of industrial technology and skilled people between industry, the universities and intermediate research institutes. In 1999, it was expanded with a focus on entrepreneurial activities and research commercialisation. In addition, the long-established Teaching Company Scheme finances an associate to work on project in a semi-academic or company environment for two years (see Box 2).

United States: The Grant Opportunities For Academic Liaison with Industry (GOALI) initiative of the National Science Foundation (NSF) funds: *i)* faculty, postdoctoral fellows and students to conduct research and gain experience with production processes in an industrial setting; *ii)* industrial scientists and engineers to bring industry's perspective and integrative skills to academia; and *iii)* interdisciplinary university-industry teams to conduct long-term projects. There are no requirements for matching funds from firms for GOALI projects performed in universities. University-industry IPR agreements must be made up front and submitted for funding consideration.

In Japan, since FY 2000 national university professors are allowed to act as board members of the Technology Licensing Organisations (TLOs), and national university employees can engage in industrial activities outside their regular work hours. National university researchers who participate in collaborative research with government research institutes can be granted leave without being penalised on their retirement allowance. The recent Mexican innovation law also facilitates researcher involvement in entrepreneurial activities.

Removal of regulatory barriers across OECD countries should foster greater researcher interaction with industry, but regulations are only one side of the equation. Interaction between researchers and industry depends heavily on incentives. Non-regulatory barriers such as faculty promotion and evaluation practices in public research institutes and universities that emphasise tenure and publishing over mobility and collaboration may act as disincentives for collaboration (see below). Consequently, many OECD countries have gone further than deregulation and have launched programmes to address disincentives to human resource-based science-industry interactions (Box 1). These programmes can be classified according to three main functional objectives:

- *Promote the training (and hiring) of students/graduates in SMEs.* Relevant programmes aim to achieve two general goals: *i)* stimulate transfer of knowledge, especially to SMEs from traditional sectors that lack the technical and financial resources to attract highly skilled graduates; *ii)* provide industry training and job opportunities for students and graduates. Support can take the form of tax credits or reimbursement of labour costs. One of the main pitfalls of such programmes is the risk of capture, whereby subsequent hiring of graduates is skewed towards those participating in the scheme. In addition, it is not always easy to ensure a satisfactory match between the skills demanded and the qualifications and research interests of graduates.

- *Promote the training of established public researchers in industry.* This is the most common approach, whereby established researchers in the public sector work with industry on specific research projects. A main lesson drawn from experience is that programmes must be sufficiently funded in order to foster lasting relations between the producers and users of knowledge. Such programmes also serve to facilitate the entry of young researchers to the labour market.

- *Encourage contact and training of industry researchers in a public research environment.* In several countries, new government and institution-based initiatives have been implemented to promote the temporary movement of industry researchers to universities, often to work on longer-term projects that would not be taken up by industry alone. The experience of the US GOALI programme highlights the importance of ensuring that IPR arrangements are settled from the outset to avoid conflict. Such schemes are effective in building formal and informal networks among researcher, thus setting the stage for future collaboration.

Mobility has a global dimension: insofar as the health of industry-science interactions depends largely on human resource interactions, OECD countries increasingly seek to attract foreign talent while facilitating the temporary circulation of nationals abroad, especially postdoctorates and researchers (Box 3).

Incentive structures: the examples of intellectual property rights and evaluation systems

Regulatory barriers and other disincentives to the industry-science linkages can significantly reduce innovation performance. While there has been much research on barriers or disincentives at the level of institutions, little attention has been paid to understanding the role of national regulations and practices in research funding, equity investments by publicly funded organisations, intellectual property rights, mobility of scientists and research personnel, and research evaluation in fostering or inhibiting industry-science linkages. The following section focuses on regulations and incentives in the areas of intellectual property rights and research evaluation.

Intellectual property rights

The ownership of intellectual property rights (IPRs) creates strong incentives for universities and public research institutions to commercialise research and knowledge. In nearly all OECD countries there has been a marked trend towards transferring ownership of publicly funded research results from the state (government) to the (public or private) agent performing the research. The underlying rationale for such change is that it increases the social rate of return on public investment in research. Where OECD countries differ is in the allocation of ownership among performing agents (research institution *versus* individual research), in licensing practices, in the allocation of the resulting royalties, and in provisions for ensuring that the nation benefits from the patentable results of public research.

Box 3. **Recent changes in policies facilitating the international mobility of highly skilled workers**

Several OECD Member countries have amended their legislation to facilitate the admission of foreign specialists, in particular in high-technology fields. In addition, countries are implementing policies to attract the return migration of nationals and the temporary international mobility of scientists.

Relaxing quantitative immigration constraints. In 2001, the United States raised the annual quota of H1B visas reserved for professionals and skilled workers by nearly 70%. Over the coming three fiscal years, 195 000 people will be able to gain temporary admission to the country under this programme. In addition, the 7% ceiling on the share of visas granted to nationals of any given country has been lifted.

Setting up special programmes for shortage occupations. In August 2000, the German Government instituted a "green card" programme under which, one year later, some 10 000 computer and information technology specialists had been brought into Germany to work for periods of up to five years.

Facilitating recruitment conditions or procedures and relaxing criteria for issuing employment visas for highly skilled workers. Since 1998, France has applied a simplified system for computer specialists, under which labour market testing is no longer required. The United Kingdom now applies simplified fast-track procedures for issuing work permits for certain occupations and has extended the list of shortage occupations. Australia has amended its points systems for permanent immigrants, giving more weight to a number of skills, including those in new technology fields. In Korea, skilled workers can now stay in the country permanently.

Allowing foreign students to change status at the end of their course and thereby enter the labour market. In the United States, almost a quarter of new recipients of H1B visas are students already residing in the country. In Germany and Switzerland, students are no longer compelled to leave at the end of their course and may apply for an employment visa. In Australia, students who apply for a temporary skilled work visa within six months of graduation are exempt from the normal requirements relating to work experience.

Improving the attractiveness of the public research sector. The UK Government plans to raise the salaries of postdocs by 25% and increase funding for the hiring of university professors. The European Commission has doubled the amount of funding devoted to human resources in the Sixth Research Framework Programme to EUR 1.8 billion with a view to raising the attractiveness of the European research area. In a joint project with the Wolfson Foundation, the UK Government is funding a Research Merit Award scheme, run by the Royal Society and worth GBP 20 million over five years. This offers additional funds to institutions to increase the salaries of researchers they wish to retain or recruit from industry or overseas. China has recently launched a project to develop 100 universities into world-class institutions that not only provide higher education training to nationals but also academic employment and research opportunities.

Providing tax incentives to encourage recruitment of foreign personnel. In 2001, Sweden passed a new law to alleviate the tax burden on foreign experts and highly skilled workers residing in Sweden for less than five years. Denmark, the Netherlands and Belgium have adopted similar policies. In Quebec, the government is offering five-year income tax holidays (credits) to attract foreign academics in IT, engineering, health science and finance to take up employment in the provincial universities.

Repatriation schemes for postdocs and scientists. The Academy of Finland has implemented a programme to ease the return to Finland of Finnish researchers who have been abroad for a length of time. In Austria, the Schroedinger scholarships aim to help returning Austrians integrate into scientific institutions. In 2001, Germany's Ministry for Research and Education (BMBF) launched a new programme to attract the return migration of German researchers. In support of the repatriation of Canadian postdoctoral researchers, the Canadian Institutes for Health Research offers a supplementary year of funding to Canadians and permanent residents who are recipients of either the Japan Society for the Promotion of Science (JSPS) Postdoctoral Fellowships for Foreign Researchers or Wellcome Trust/CIHR Postdoctoral Fellowships. In order to be eligible for the "Canada Year" funding, training must take place in a Canadian laboratory.

Source: OECD (2002), *International Mobility of the Highly Skilled.*

How do OECD countries compare in the allocation of ownership? In the United States, the well-documented Bayh-Dole Patent and Trademark Amendments Act of 1980 allowed performers of federally funded research to file patents on the results of research and to grant licences for these patents to third parties. In addition, the 1980 Stevenson Wydler Innovation Act (amended in 1986 by the Federal Technology Transfer Act) authorised federal laboratories to conduct co-operative research and development agreements (CRADAs) with private firms and to assign any resulting patents to these firms. The Bayh-Dole Act is being emulated across the OECD area and beyond. Several European OECD countries have recently reviewed or modified ownership rules, notably Denmark in 2000 and Germany in 2001. However, a few countries grant ownership to the inventor (Table 7). For example, in contrast to the general trend, the Italian Government implemented legislation in 2001 granting IP ownership to researchers at universities. While this may provide commercialisation incentives to individuals, it may raise the transaction costs associated with collaborative research involving, on the one hand, firms and, on the other, foreign universities where ownership is retained at the institution level. In fact, in several countries there is a high level of heterogeneity in the allocation of title from publicly funded research, which adds to the complex web of regulations governing co-operation between the public research sector and industry, but also among public research institutions themselves. For example:

- In Canada, about half of the universities grant ownership to individual researchers with the remainder retaining title.

- In Finland, researchers at university – but not at the Academy of Science – own the IPRs from their inventions.

- In Germany, public research institutes receiving public grants own the resulting IPRs. In universities, a new law, which enters into force in 2002, will shift title to invention from the professor to the university and will make provisions in support of patenting and licensing activities and for the sharing of royalties equally between the university, the commercialisation unit and the professor (inventor).

- In the United Kingdom, there are also various sets of rules, although there is a move to shift title to the institution, in particular in the new polytechnics. The UK's Biotechnology and Biological Sciences Research Council's (BBSRC), one of the government-funded research councils, grants title to IPRs to the individual institutes, while the Medical Research Council retains IPRs ownership itself. Each of the BBSRC institutes is permitted to retain exploitation income equivalent to up to 10% of its annual recurrent income from BBSRC. If revenues exceed the 10%, the institute may make a proposal for reinvestment.

- In France, the ownership of IPRs at PROs is generally granted to institutions, but in practice this varies widely depending on the institution. Traditionally, the national scientific research centre (CNRS) allows industrial partners to own the patent but negotiates royalties in exchange. In the life and agricultural sciences, the INRA and INSERM institutes own the patent but often grant exclusive licences to partners. At universities, ownership in principle belongs to the institution but the complex nature of public research in France wherein universities and public labs jointly perform government-funded research means that in practice ownership could be shared.

Table 7. **Ownership of IPRs in publicly financed research organisations (PROs)**

	Ownership of IPRs		
	PRO	Inventor	Government
Australia	◆		
Austria[1]	◆ (P)	◆ (U)	
Belgium	◆		
Canada	◆	◆	
Denmark[2]	◆		
Finland[3]	◆ (P)	◆ (U)	
France	◆	◆	
Germany	◆		
Hungary	◆ (P)		
Iceland	◆ (P)	◆ (U)	
Italy	◆ (P)	◆ (U)	
Japan		◆	◆
Korea[4]			
Mexico	◆ (P)		
Netherlands	◆ (P)		
Norway	◆ (P)	◆ (U)	
Poland	◆		
United Kingdom	◆		
United States	◆		

(P) = Non-university PROs (public labs, academies, etc.).
(U) = Universities.
1. In Austria, the government owns inventions by employees at universities, but in practice transfers ownership to the individual inventors.
2. In Denmark, ownership is claimed by the university or PRO but inventors have a right of first refusal.
3. In Finland, ownership of inventions at non-university public research organisations must be transferred from the individual to the institution, provided the latter can exploit the invention.
4. In Korea, ownership is dependent on the research contract and source of funding. Researchers at both universities and PROs own title to invention for privately funded research.
Source: OECD.

- In Japan, while in principle the government owns intellectual property resulting from research performed at national research institutes, joint IPR ownership between the government and researchers is now being permitted. Nevertheless, governmental consent continues to be required for the exploitation of IP by third parties, which may slow the commercialisation process. In universities, title to an invention is determined by a committee, which frequently decides in favour of the researchers themselves. In 1998, invention committees awarded ownership to researchers in 77.9% of cases (Shimoda and Goto, see Chapter 6). New initiatives have been taken to better support the commercialisation of inventions held by individual researchers. Companies which contract research are now allowed to own a part (up to half) of the rights to patents originating from contract research. With regard to collaborative research, since 1999 the ownership of results of public research performed by a private firm (but paid for by the government) can belong to the firm in exchange for royalty-free licensing to the government.

The lack of clarity and diversity in national and institutional guidelines for IPRs can be a barrier to commercialisation insofar as it increases the risk and transaction costs of co-operation for industry, especially for SMEs, which often lack information and experience in accessing public research. For example, the Canadian Government is reviewing practices and assessing the need to promote harmonisation. Harmonisation efforts raise the question of best practice regarding IPR ownership.

Is granting ownership to the researcher a good formula? In theory, it should increase researchers' interest in commercialisation. However, putting all the responsibility for disclosing and protecting ownership on a single individual reduces the likelihood of patenting and subsequent licensing.[3] In addition, the growing costs of litigation act as a powerful disincentive for disclosing and commercialising research results when IPRs are owned by individuals. Firms, too, may hesitate to enter into agreements where there is a risk of future litigation from one or more of the co-inventors. Another problem is that the researcher who owns the IPR may take it abroad for commercialisation, thus reducing national benefits from public investment. Given these drawbacks, a good practice might be to grant IPR ownership to the research organisation but to ensure that researchers enjoy a fair share of any resulting royalties.

How to share royalty revenues? In general, the main beneficiary is the owner of title. However, sharing of royalty revenues is common across countries and institutions, and is increasingly seen as a way to provide incentives not just to individual researchers but to research teams. There are generally no standard national formulae for allocating royalties from patents and licences, but national guidelines do matter.

- The United States' Bayh-Dole Act, for example, stipulates that royalties from licensing be shared with the inventors and that remaining income, less payment of expenses, be used to support research and education within the university. A common practice in US universities is to grant royalties subject to meeting a certain revenue threshold. For example, royalties may only be paid on licences that generate significant revenue above the university expenses incurred for patenting.

- In most countries (Austria and the Netherlands are notable exceptions) net royalty revenues are shared between the institution and the inventing researchers and, in some cases, the division or the unit where the research was conducted also benefit (Table 8).

- In some cases, a greater share is granted to the individual researcher, such as in the University of California system, where inventors enjoyed 40% of net royalty revenues in 1997. At Warwick University in the United Kingdom, academics can receive 75% of royalties up to a certain threshold, after which the share drops to 50% for the researcher and 50% for the university. In France, inventors at universities are granted 50% of royalties paid to institutions. Current discussions are aimed at modifying the arrangements for the sharing of royalties so that other individuals or research teams that contributed in some manner or another to the invention can be rewarded. In Japan, the Japan Science and Technology Corporation (JST), to which university inventions are transferred, bears the costs of patent application and renewal. If the commercialisation is successful, JST returns 80% of royalties to the inventors (researchers).

In multidisciplinary research organisations such as universities, there is a potential for tension insofar as high-yielding patents and licences tend to be concentrated in specific technological fields (*e.g.* biomedical). The question arises whether the royalties should be shared not only among the main agents behind the invention but also with other departments. This could partly answer the concern that those research areas that generate the bulk of external revenues could attract also an increasing share of intramural funding, to the detriment of other disciplines (*e.g.* physical sciences, social sciences and humanities).

Table 8. **Guidelines for sharing royalties from IPRs**

	Apply to	Share of royalties			
		Inventor	Laboratory/department	Institution	No sharing
Australia	Universities	33%	33%	33%	
Austria	General practice				100% to owner
Belgium	Flemish universities	10 to 30%	50%	20 to 30%	
Canada	Federal research	35% by law	variable	variable	
France	Public labs	25%	25%	50%	
Germany	Max Planck and HGF centres	33%	33%	33%	
Hungary		0%	undetermined	up to 100%	
Israel	Hebrew University	33%	33%	33%	
	Weizmann Institute	40%	0%	60%	
Japan	Universities				100% to owner
Korea	KIST institute	up to 60%	0%	40%	
Mexico	Public labs				100% to owner
Netherlands	Public labs				100% to owner
Poland		no general rule			
United Kingdom	BBRCs	sharing encouraged in institute guidelines			
United States	Universities	sharing required by law			
	Stanford	33%	33%	33%	

Source: OECD.

What should be the role of government in guiding the licensing policies of publicly funded research organisations? While the decision to licence on an exclusive or non-exclusive basis generally devolves to the titleholder, governments do play a role. First, granting ownership rights to institutions rather than to researchers encourages more non-exclusive licensing. Governments also influence licensing options by helping to define what can and what cannot be patented and, finally, by providing the infrastructure for licensing. Publicly funded research organisations may be encouraged to favour non-exclusive, but royalty-bearing, licences on the grounds that this ensures a broader diffusion of knowledge. In addition, such a practice does not entail restrictions on the freedom to publish, a major issue with exclusive licences in some fields. One study found that nearly three-quarters of the active licences granted by six of the largest US research funding agencies (*e.g.* the National Institutes of Health, the Department of Energy) were non-exclusive during the fiscal years 1996-98 (GAO, 1999). However, the share of exclusive licences was significantly higher in the portfolio of research-performing organisations, reflecting the fact that firms, particularly in sectors where product development is very capital intensive and lengthy, often require exclusive rights.

Table 9. **Exclusive *vs.* non-exclusive licensing of patents from public research**

	Exclusive	Non-exclusive
	For public research	
Advantages	o Speeds technology transfer o Effective in attracting investors, especially for SMEs and spin-offs	o Fosters broader diffusion o Broader revenue base from royalties o Reduces risk of conflict of interest
Disadvantages	o May limit diffusion of knowledge o Raises obstacles to research requiring patented knowledge o Review process may be slow o Risk of litigation	o Requires more resources to manage and advertise licensing opportunities
	For firms	
Advantages	o Reduces development risk o Generates monopoly returns	o Larger firms benefit from market power
Disadvantages	o Small firms may be disadvantaged o Higher share of royalty	o Competitors may develop technology first

Source: OECD.

How to ensure national economic benefits? In many countries, regulations governing public funding of industry-science partnerships or collaborative R&D programmes and licensing of the resultant IPR to foreign partners are subject to restrictions to ensure national economic benefits. In the United States, participants in CRADA agreements must show that benefits will accrue to the United States, usually by agreeing that commercial production will take place in the United States. Australia's Research Council programmes allow foreign participation but require that the IPRs are used in such a way as to maximise benefits to Australia. In Canada, there is concern about possible leakage of IPR knowledge to foreign countries, notably the United States, via the mobility of researchers who, as owners of IPR, can commercialise their research results abroad. A general problem with rules on national economic benefits is that such requirements tend to be interpreted very differently by the different stakeholders.

Building the infrastructure and skills base for IP management. Generally, in European countries, public research organisations, such as Germany's Max Planck and Fraunhofer Society, and France's public labs (*e.g.* INRIA, INRA), have a more developed institutional infrastructure for commercialising IP than do the universities. An important challenge facing universities in a number of countries is not only the disincentives from legislative arrangements relating to IP ownership, but a lack of capacity in the skills base for managing IP. In France, for example, although commercialisation offices at universities have been existence for some time, their mission has been more oriented towards administrating the IP related to contract research with firms rather than to encouraging inventors to disclose and patent new inventions or marketing university inventions to firms. Shifting the emphasis to the latter will require investing in human capital that is scientifically trained to earn the respect of researchers but entrepreneurial enough to collaborate with industry. In the United Kingdom, the government plans to commit GBP 10 million (EUR 16.7 million) to strengthening the capacity for commercialising IP at the Public Scientific Research Establishments.

National patent offices and university licensing associations, such as the US Association of University Technology Managers (AUTM), the UK's Association of University Research and Innovation Links (AURIL) and the Association of European Science and Technology Transfer Professionals (ASTP), can play an important role in improving the patenting capacity and skills bases of universities by increasing awareness and sharing best practices. The UK Patent Office has undertaken to reduce charges for patenting. In partnership with the AURIL, it also will develop a Web site to increase awareness about patenting among universities. The French Patent Office has also launched information campaigns to encourage patenting by researchers, entrepreneurs and SMEs. In Japan, the 1998 Technology Transfer Law grants subsidies and reduces patent fees for technology licensing organisations (TLOs) that meet certain criteria and have been approved by the Ministry of Education and the Ministry of International Trade and Industry.

Research evaluation systems

Public research institutions are being asked to contribute to economic development but also to be more responsive to evolving societal concerns such as food safety, environmental degradation, and health issues. These pressures for greater accountability are in some countries the counterpart of greater autonomy, and everywhere they encounter strong resistance from the research community which fears that, under the cover of noble motives, changes in evaluation criteria could reduce core funding and/or shift it away from longer-term free research. What is put into question is both the sole focus on scientific excellence and the criteria for judging this excellence when evaluating public researchers and research institutions. Evaluation of research must evolve for at least reasons. First, its scope must be broaden in response to the considerable expansion of the commercialisation activities of universities and public research institutes (*e.g.* licensing offices, venture funds, spin-offs). Second, evaluation criteria must take into account that excellence in research and training of graduates has become, at least in some disciplines, more tied to applications in industry.

In the case of *applied research institutions* (*e.g.* CSIRO in Australia, Fraunhofer in Germany), countries have generally chosen to retain traditional criteria (peer review and publications) when evaluating research eligible for core funding, but have made core funding increasingly dependent on the level of industry financing, thus implicitly changing overall evaluation criteria. Some have also included "commercialisation clauses" in competitive research grants.

- External financing targets vary from 20% to 50% (*i.e.* matching funds). In addition to requirements for external industry financing, in some cases evaluations integrate input and output measures of commercialisation, such as the amount of internal R&D funds invested in collaborative R&D projects, income from contract research, and the number of patents, joint publications and inventions, the amount of licensing income, etc. In Korea, for example, the 1999 reform of the public R&D system requires the evaluation of the Government Research Institutes (GRIs) to consider, *inter alia*, the amount of co-operative R&D and the royalty and licensing income from SMEs, as well as the number of spin-offs generated. The performance of the GRIs relative to these criteria is considered in the annual budgeting process. The individual institutes are encouraged to develop their own criteria by which researchers are evaluated relative to the institute's overall performance. One of the secondary objectives is that GRI researchers become more customer focused.

- Germany has instituted new rules to BMBF grant-awarding practice that require that those who obtain research results through BMBF funding must apply for IPRs and commercialise the results but in exchange gain exclusive title to IPRs, including the income from licensing.

Box 4. **Showcase on evaluation of research and researchers:
the Flemish Inter-University Institute for Biotechnology (VIB)**

Institution

The Flemish Inter-university Institute for Biotechnology (VIB), founded in 1995 by the regional government of Flanders, combines nine university departments and five associated laboratories, representing over 700 researchers. The VIB has three major objectives: *i)* to perform quality research; *ii)* to foster technology transfer through licensing and spin-offs; and *iii)* to enhance the public image of biotechnology.

Evaluation of scientific research

Annual funding is contingent on evaluation with regard to scientific quality and commercialisation. In the scientific evaluation, each department is ranked (1, very poor; 2, very good; 3, excellent), with poorly ranked departments being discontinued. To complement the long-term four-year evaluation, the Institute is considering using the Science Citation Index as a continuous monitoring tool. However, care is taken to ensure that researchers do not place too much emphasis on the short-term publication agenda to the detriment of research. The VIB's research groups are ranked above average in the Science Citation Index (SCI/budget) and the average number of lines of invention and patent applications is increasing each year. With regard to lines of invention, the Institute was classified in the top 10% of all research departments operating on similar budgets.

Evaluation of technology transfer and commercialisation

To evaluate technology transfer and commercialisation, the record of inventions – a standard measure in the United States – is used as a monitoring instrument. Research groups must disclose each invention or potential invention to the technology transfer group of VIB. Comparable universities or institutes are chosen as benchmarks. VIB sees this as a very important evaluation criterion for its research departments. In addition, VIB takes into account the number of research collaborations and licence agreements of each research department, although this is of lesser importance than the record of inventions. The goal is that the research departments should excel in research, leaving VIB to commercialise the technology.

Recruitment and retention policy

Recruitment at VIB is organised according to occupations within departments and, for scientists, track record and publications are the most important elements. VIB makes a distinction between scientific researchers and scientific personnel in management positions. This contrasts with the flat structure in university departments where a professor oversees a team of research scientists. The additional management layer, which is based on the Cambridge model, guarantees a career for the scientific middle management. In an effort to achieve a balance between short-term rotation and long-term stability, all key researchers are given a long-term employment contract.

Elements of best practice

The main strength of the VIB model is excellence in the interface department (commercialisation unit), which offers business and technical incubation services. VIB expects its researchers to perform world-class research and not to get engaged in real commercialisation activities. Its internal evaluation system is mainly geared towards providing incentives to perform leading research. In contrast, the commercialisation unit focuses almost exclusively on innovations with potential commercial application. If a line of research can be patented, VIB applies for a patent. Whenever possible, a spin-off is created (*i.e.* five spin-offs have been created, of which three in the last year). However, the inventor is not necessarily the entrepreneur in the spin-off.

Different approaches are needed for balancing incentives for commercialisation with support for longer-term research in *universities and basic research institutes*:

- New Zealand's Foundation for Research, Science and Technology (FRST) funds research at Crown Research Institutes and universities that lead to societal and economic benefits (*e.g.* outputs), but not necessarily commercialisation activities *per se*. The Foundation assesses research projects on the basis of merit with respect to long-term outcomes, the track record of the research team, and linkages with industrial partners as a measure of future commercialisation. A distinction is made in the evaluation criteria between projects in fundamental research and those in applied research. In addition, by granting IPR ownership to the institutions performing the research, incentives are built in for the institutions to establish linkages with industry where appropriate.

- Another possibility is to separate funding for commercialisation activities from core intramural research funding. The Japanese Ministry of Education has created a budgeting scheme whereby national universities promoting university-industry co-operation and patenting can be allocated additional funds. In France, financial incentives to research have been increased and are channelled through two distinct funds: one to support technological development; the other to promote more fundamental research.

- Other OECD countries are planning significant changes in the financing of university research. For example, Australia is leaning towards reforming its university research financing system to put in place better incentive structures for universities to draw research funds from industry.

Rewarding individual researchers for their contributions to commercialisation goals is another means of improving linkages between public research and industry. However, this approach is still under-exploited in the majority of countries. In the United Kingdom, the Realising Our Potential Award Scheme rewards academic researchers who obtain financial support from industry by granting them award money to pursue curiosity-driven research. New Zealand's Crown Research Institutes provide specific awards for the development of IPR and commercialisation of research outputs. In the United States, there are proposals to develop a national innovation award along the lines of the national awards in science and technology to recognise commercialisation. In Mexico, the recent innovation law broadens the eligibility criteria of the "national system of researchers" (SNI) in favour of top-rate researchers involved in applied research and technological development.

Institutional arrangements

Changing incentive structures induce some institutional changes (*e.g.* the proliferation of technology transfer and licensing offices at universities following the Bayh-Dole Act in the United States), but may require others to yield their full benefits. This concerns the overall institutional profile of national systems of ISRs (Table 10) and the organisational framework of commercialisation activities at universities and public laboratories.

Overall institutional profiles

While the assumption underlying M. Porter's innovation index (Porter and Stern, 1999) – that the degree of "connectivity" within the innovation system is a function of the share of universities (and thus in inverse relation with that of public institutes) in government-sponsored R&D performance – is highly simplistic, it does point to a real issue of concern, although it overlooks a number of others:

Table 10. **ISR-related institutional settings in selected countries at the end of the 1990s**

		Austria	Belgium	Finland	Germany	Ireland	Italy	Sweden	UK	US	Japan
Institutional structure of public research *(share in total R&D in public research in %, partly estimates)*	Universities (incl. technical univ., univ. of arts, univ. of theology, other specialised univ.)	76	82	59	52	61	53	82	58	64	58
	Polytechnics and HE colleges	~1	~2	~1	2	5	~1	4	~1	-	3
	Primarily transfer-oriented PSREs		10	18	~7	~12		14	19	24	
	Large research centres with strategic mission	11	-	-	16	-	~22	-			
	PSREs specialised on basic research	8	3		~11	-	~24	-	12	5	39
	Departmental PSREs, others	4	3	22	12	~22		-	10	7	
Governance of public research	Competition-based R&D financing in HEIs	☐	■	■	■	■	☐	■	■	■	☐
	Competition-based R&D financing at PSREs	■	■	■	■	■	■	■	■	■	☐
	Third mission of universities	☐	☐	■	☐	■	☐	☐	☐	■	☐
	Technology transfer as part of evaluation in HEIs	☐	☐	☐	☐	☐	☐	☐	☐	■	☐
	Relevance of private HEIs	☐	☐	☐	■	☐	☐	■	■	■	■
	Thematically specialised PSREs with transfer mission	■	■	■	■	■	■	■	■	■	☐
	Industry representatives in advisory boards, etc., of PSREs	■	■	■	■	☐	■	■	☐	■	☐
Intermediary infrastructure	Technology transfer offices in HEIs	■	■	■	■	■	■	■	■	■	■
	Commercialisation enterprises, transfer institutes in HEIs	☐	■	■	■	■	■	■	■	■	■
	Science parks and incubators in HEIs	☐	☐	■	■	■	☐	■	■	■	■
	Intermediaries at the level of industry associations etc.	☐	■	☐	■	■	☐	■	■	■	☐
	(Semi-)public technology and innovation consultants for SMEs	■	■	■	■	■	■	■	■	■	■
	Regional consulting networks	☐	■	■	■	■	■	■	■	☐	■
	Information service provision for technology transfer	■	■	■	■	☐	☐	■	■	■	■
	Significance of private intermediaries	☐	■	☐	■	☐	☐	■	■	■	☐
	Joint industrial research networks at sector level	■	■	☐	■	☐	☐	■	☐	■	■

Note: ■ : high significance; ■ : important; ☐ : less important/missing.
Source: Austrian Federal of Economy and Labour (2001).

Box 5. Benchmarking indicators

Institution-level benchmarking indicators
The example of Risö* (Denmark)

	Departments	Programmes
Activity indicators	o Own research and development Co-financed contracts o Client-financed contracts Technical and other support to research o Programme management Tasks, contracts o Tasks, others Management of department Total sum, man/months	o Own research and development Co-financed contracts o Client-financed contracts Management and administration o Programme management Total sum, man/months Personnel, man/years
Production indicators	o International publications Of which, reports o Danish publications Conference contribution to proceedings o Popular scientific articles Personnel, man/years (PhD & post doc)	o International publications Of which, reports o Danish publications Conference contribution to proceedings o Popular scientific articles Patent application
Network indicators	o Number of completed PhD projects Expatriates (min. 1 month), months o Guest researchers (min. 1 week), months Co-operation with companies, months o Companies engagement, months Co-operation with research org., months o Research org. engagement, months Co-operation with government o Patent applications	o Number of completed PhD projects Expatriates (min. 1 month), months o Guest researchers (min. 1 week), months Co-operation with companies, months o Companies engagement, months Co-operation with research org., months o Research org. engagement, months Co-operation with government

* Risö is the main public research institute in Denmark.

Source: "Three year plan 1998-2001", Risö, as quoted by ADL (2000).

Country-level benchmarking indicators
Comparison of commercial outputs of Australian and US publicly funded research

Benchmark	In 1996, for evry 1000 staff an average US university received 88 million $ grants from industry and 320000 $ in royalties, made 9 patent applications, negociated 11 licenses, and generated 1 spin-off (Trune, 1996*)			
Australian performance (1996/97)	**Multidisciplinary research**		**Medical research**	
	Actual	Expected	Actual	Expected
Industry funding (million USD)	60.3	49.3	6.7	9.1
Patent applications	52	60	n.a.	27
Royalties (million USD)	4.0	2.2	n.a.	0.96
Spin-offs	2.5	7	n.a.	1.3

* Trune, D. (1996), "Comparative Measures of University Licensing Activities", *Journal of the AUTM*, vol VIII.

Source: OECD, based on Thorburn (1999).

- *Do university-based systems of ISRs enjoy a comparative advantage?* Yes probably, especially as ISRs increasingly require multidisciplinarity and build on people-based interactions – but with some caveats. The United States and the United Kingdom demonstrate that universities can provide excellent platforms for vibrant industry-science relations. However, there are also counter examples, such as Japan where universities are central in poorly performing ISRs, or Australia where the commercialisation activities of the major public institute (CSIRO) compare favourably with those of an average US university (Box 5).

- *Under what conditions can increasing the share of universities in publicly funded R&D performance improve ISRs?* In the last decade, a majority of OECD countries have redirected public R&D investment towards universities, to the detriment of research institutes. However, this shift has not improved ISRs to the same extent everywhere. A major reason is that decentralised university systems, in which universities enjoy more freedom in their research policy and relations with industry, are more responsive than centralised ones to opportunities for ISRs. Although the latter may be justified on other grounds, governments should realise that they have are increasingly costly in terms of reduced commercialisation potential. Governments should also accept the fact that giving more weight to commercialisation objectives in allocating core funding will only serve to accentuate the polarisation of university research capabilities around existing centres of excellence.

- *How to enhance the contribution of public research institutes to innovation without distorting the market for contract research and technological services?* Although declining in recent years, the role of these institutes within ISRs remain important in most OECD countries. Their size, missions and research portfolios are extremely diverse, even when judged from a limited sample (Table 11). However, improving their linkages with industry poses some common generic problems and has become a priority in some countries (Box 6). First, their ageing staff have lower incentives to co-operate with younger researchers in industry, and the downsizing of operations has exacerbated this "generation gap" in recent years. Second, the objectives of government and those of the research organisations may be not consistent, especially when public institutes have a strong applied research focus and are under pressure to downsize. Governments wishing to maximise the economic spillovers from research run the risk of encouraging diversification away from core missions and subsidising the development of market-distorting technological services.

- *How to increase synergies between different publicly funded research organisations to the benefit of commercialisation and innovation? What role for government in intermediation?* Different countries have chosen different approaches that would deserve a thorough comparative evaluation. One is to develop organic ties between universities and public institutes (*e.g.* CNRS labs within universities in France). A second approach is to add an institutional layer that provides a stable meeting place for collaborative research (co-operative research centres in Australia and Austria). A third approach uses catalytic programmes (*e.g.* thematic research networks or programmes in France, Japan, the Netherlands and the United States). A fourth approach is to change the mission of existing technology transfer organisations (*e.g.* Fraunhofer in Germany) or to create new types of intermediaries specialised in IPR transactions (see below).

Governments generally lack organised information and appropriate review mechanisms to answer some of these questions. They should reinforce the co-ordination of commercialisation activities by publicly funded research organisations, as well as their monitoring and evaluation, including through benchmarking against international best practice (ADL, 2000).

		CNRS (France)	**CNR** (Italy)	**CSIC** (Spain)	**CSIRO** (Australia)	**Centre Juelich (FZJ)** (Germany)	**Lawrence Berkeley** (United States)
P r o f i l e		Multidisciplinary basic research centres					
	Mission	The main research centre in the country				One of the largest centres of the HGF Association	One of the DOE (Department of Energy) research laboratories
	Staff	25 400 (11 470 researchers)	7 500 (3 700 researchers)	9 000 (2 345 senior researchers)	6 700	4 300	3 800
	Budget (98)	EUR 2.4 billion	EUR 698 million	EUR 340 million	AUD 730 million	EUR 230 million	USD 340 million
F u n d i n g	Institutional	90.5% (1998)	76% (1998)	60% (1998)	65.3% (1998/99)		
	Gov. contracts & competitive grants	9.5% (1998)	19% (1998)	n.a.	21.6% (1996/97)	n.a.	n.a.
	Industry		5% (1998)	6.2% (research contracts) (1998)	11% (1996/97)		
I n d i c a t o r s	Contracts and joint labs with industry	3 000 contracts, 26 joint labs	447 contracts	505 contracts, 1 joint lab	Participates in 51 of the 65 Co-operative Research Centres	n.a.	Entered into 140 CRADAs during the 1990s
	Inventions, patents, licences	Stock of 4 181 patents (of which around 50% co-owned) and 500 licences	Stock of 550 patents (36 new patent applications) and 95 licences	66 new patent applications (1998), stock of 600 patents and 210 licences	51 new patent applications (1997/98)	Stock of 607 national and 3 944 international patents, and 169 licences	3 new patents issued (1998), 56 new licences (1997), stock of 16 patents
	Licensing revenues	EUR 29 million (2000)	EUR 0.35 million (1999)	n.a.	AUD 5.26 million (1997/98)	EUR 3.6 million (1998)	USD 0.5 million (1997)
	Spin-offs	221 since 1985	n.a.	n.a.	4 in 1999, 56 since 1972	6 in 1998, 26 since 1983	n.a.
	Organisation of knowledge-transfer activities	A central unit (DAE) evaluates potential, defines IPR strategy and negociates projects and royalties	A Technology Transfer Office (DAST) deals with all issues related to technology transfer	A Technology Transfer Office (OTT) is in charge of fostering and managing all activities in conjunction with industry	Quite decentralised, with support from the Corporate Business	A Technology Transfer Office (TTB) deals with all issues related to technology	A Technology Transfer Department (TTD) deals with all issues related to
	IPR management	A major part is subcontracted to a specialised affiliate, FIST	Performed by DAST	Performed by OTT	Performed by CBD, under supervision of the Intellectual Property Standing Committee	Performed by TTB	Performed by TTD with support from the Office of Technology Transfer of the University of California
	Start-up policy	Low but increasing focus; DAE conselling support to PhD students; initiation of a nation-wide network of local incubators in partnerships with other research institutes	Low focus	Low focus	Low but increasing focus; different models of spin-off formation, with equity participation in 16 of the 56 spin-offs formed since 1972	Low but increasing focus; TTB provides some financial and administrative support, takes part in a regional initiative to support entrepreneurs and has a stake in the "Technologiepark Jülich"	Low focus

Source: OECD, based on ADL (2000) and other sources.

		Fraunhofer Gesellschaft (Germany)	**INRIA** (France)	**INSERM** (France)	**Massachussets General Hospital** (United States)	**Independant MRIs** (Australia)	**DERA** (United Kingdom)
P r o f i l e	Mission	Multidisciplinary applied research	Thematic research centres				
		The main German applied research institute (federates 48 centres)	Research on IT	Medical research			Defence research
			The main French public institute for research on IT	Federates over 250 labs in hospitals and universities	The largest hospital-based research centre in the US	Institutes that are not departments of a hospital or university	The Agency in charge of most non-nuclear R&D
	Staff	9 000 (3 000 researchers)	2 100 (715 permanent staff)	10 000 (2 970 researchers)	n.a.	3 000	11 500 (1 000 PhDs)
	Budget (98)	EUR 665 million	EUR 75 million	EUR 460 million	USD 200 millions	AUD 130 million	EUR 1.5 billion
F u n d i n g	Institutional	70% (average over last 5 years)	n.a.	89% (1998)	n.a.	n.a.	92% (1998/99), of which 90% from Ministry of Defence (MOD)
	Gov. contracts & competitive grants			5.6 % (1998)			
	Industry	30% (average over last 5 years)		5.4 % (1998)	18% (1998)	17.5% (sample of 5 MRIs)	8% (1998/99)
I n d i c a t o r s	Contracts and joint labs with industry	n.a.	300 contracts, 4 co-operative ventures	n.a.	n.a.	n.a.	Sub-contracts 32% of all its MOD funded research
	Inventions, patents, licenses	417 patent applications, 90 new licences (1998)	n.a.	Stock of 331 national & 1 262 international patents, and 253 licences	145 invention disclosures, 140 patent applications and 57 new patents issued (1998)	n.a.	111 patent applications, 68 new licences (1998). Stock of 6 000 patents and 500 licences
	Licensing revenues	EUR 3.0 million (1998)	n.a.	EUR 8.9 million (1998)	USD 1.8 million (1998)	n.a.	n.a.
	Spin-offs	n.a.	5 in 1998, 40 since origin	15 since origin	3 in 1998	2 in 1999, 11 since origin	n.a.
	Organisation of knowledge transfer activities	Decentralised, with support from the Fraunhofer Patent Centre (FPC)	A specialised Department (DirDRI) manages technology tranfers and assists research teams in their relations with industry	A specialised Department (DPES) manages technology tranfers and partnerships with private firms	Quite decentralised, under the supervision of the Committee on Industrial Relations and Intellectual Property	Arrangements vary from one institute to another	A DERA Office (DERAtech) manages commercialisation activities, and an outside Agency (DDA) promotes access by SMEs to DERA expertise
	IPR management	Performed by the Patent Department of the FPC	Performed by DirDRI	Performed by a specialised unit (PDE) of DPES	Performed by the Office of Corporate Sponsored Research and Licensing	Arrangements vary from one institute to another	Performed by DERAtech
	Start-up policy	Low focus	Important focus; a subsidiary, INRIA-Transfert (created in 1998) supports spin-off creation and holds 34% share in I-Source gestion, a dedicated seed-capital fund	Low focus; different models of spin-off formation, with equity participation in 4 of the 40 spin-offs formed since origin	Low focus	Low focus	Low but increasing focus

Source: OECD, based on ADL (2000) and other sources.

Box 6. Reforming public laboratories in Norway and Germany

Norway

Public laboratories have traditionally played an important role in Norwegian R&D policy. In 1997, these institutes performed 26% of all R&D conducted in Norway. Until the 1980s all publicly funded technological and industrial research was carried out in designated public labs owned by the Norwegian Research Council for technical-industrial research (NTNF). However, it became increasingly clear that the dual function of priority setting for R&D support and management of a large number of research organisations was an undesirable model.

It was decided in the early 1980s to separate the two functions to create a more decentralised system. The public labs were transformed into market-oriented research institutes that received a combination of basic institutional funding and programme funding.

In the late 1980s, the system of funding the institutes was modified. Only industrial firms could apply for project funding under the new model, called user-oriented management of R&D. The funding system underwent other transformations and now consists of three parts, of which the first two account for 10-20% of an institute's typical turnover: basic institutional funding; strategic, institute-level programmes; and competitive programme, or project funding with industrial partners.

The successive changes over the past 20 years have led to a better division of tasks and responsibilities between the public agencies responsible for strategic task and priority setting and the labs.

Germany

Germany established 16 public labs* between 1956 and 1992. These employed a staff of 23 000 people in 2001 and received approximately DEM 3 billion a year in institutional funding – equivalent to 25% of all public R&D funding.

The laboratories have been criticised for the lack of co-operation among institutions and a lack of flexibility in their research approaches. A recent evaluation of the public labs showed that their potential and resources were not being used efficiently. A proposal has therefore been made to gradually move away from an institutional funding to a programme-oriented funding. The objective of the new funding system is to allocate resources along the lines of thematic research programmes that cross institutional boundaries and on the basis of an external evaluation according to international standards.

In the proposed system, the government would set priorities for the programmes to be funded after consultation with the science community, the business sector and the labs concerned. Programme portfolios, running over several years and defining clear interim milestones, as well as the share of work and budget of the institutions involved, would be established for individual research subjects. Research proposals submitted on this basis would be evaluated *ex ante* by an international evaluation team. The government anticipates that this reform would produce several benefits:

- More focused allocation of R&D funds, with greater transparency in regard to priority setting, selection of research proposals and allocation of funds.
- Improved planning security due to the fixed term of the programmes.
- Greater competition for resource allocation tempered by increased networking between institutions and improved international collaboration.
- Strengthening of scientific excellence, promotion of interdisciplinary research and co-operative research approaches with industry.

---

* Public labs in Germany (*Grossforschungseinrichtungen*) are research institutions outside the universities which are jointly funded by the Federal government (90%) and the *Länder* governments (10%).

Institutional arrangements for the management of IPR from public research

How can commercialisation activities in the public research sector be organised, taking account of the need to minimise conflicts of interest while providing the efficient legal and managerial support for protecting and licensing IPR or carrying out spin-off activities? What should be the model for dedicated institutions? Should they be located on or off campus or in public laboratories? Various approaches have been adopted in the OECD countries and these can be summarised in three main institutional models (Figure 8).

Figure 8. Organisation of ISRs in publicly funded organisations

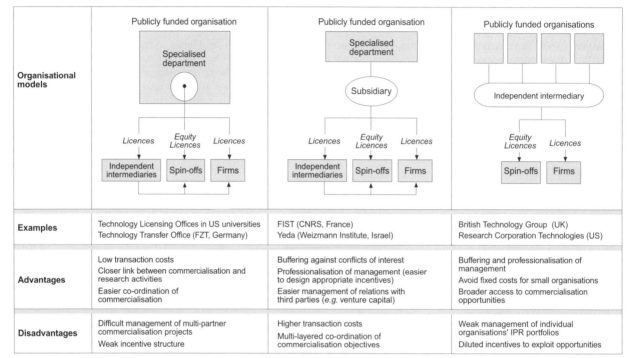

	Publicly funded organisation	Publicly funded organisation	Publicly funded organisations
Examples	Technology Licensing Offices in US universities Technology Transfer Office (FZT, Germany)	FIST (CNRS, France) Yeda (Weizmann Institute, Israel)	British Technology Group (UK) Research Corporation Technologies (US)
Advantages	Low transaction costs Closer link between commercialisation and research activities Easier co-ordination of commercialisation	Buffering against conflicts of interest Professionalisation of management (easier to design appropriate incentives) Easier management of relations with third parties (*e.g.* venture capital)	Buffering and professionalisation of management Avoid fixed costs for small organisations Broader access to commercialisation opportunities
Disadvantages	Difficult management of multi-partner commercialisation projects Weak incentive structure	Higher transaction costs Multi-layered co-ordination of commercialisation objectives	Weak management of individual organisations' IPR portfolios Diluted incentives to exploit opportunities

Source: OECD.

- *Technology transfer and licensing offices may be integrated as part of the research institution.* On-site departments benefit from low overheads as the fixed costs for staff and operations are absorbed by the main institutions. Physical presence also ensures closer links between commercialisation and research activities. However, there is the risk that on-site agencies focus mainly on existing relationships to the neglect of new opportunities. There is also an issue of scale, as smaller universities often lack the resources and technical skills to effectively staff on-site licensing agencies and their patent portfolios may be too small to justify such organisational investment.

- The advantages of *the arm's-length subsidiary* approach is that it provides a greater "buffer" against possible conflicts of interest between the commercialisation operations and the research activities. Subsidiaries are characterised by a higher degree of financial and managerial independence that facilitates relations with venture capitalists and potential licensees. Indeed, evidence at the institution level suggests that success in licensing depends heavily on relations with other knowledge production centres and geographic proximity to client firms and financial capital. Increasingly, following the pioneering example of Stanford, technology transfer and licensing offices are moving away from acting as purely legal support units to become entrepreneurial agents, not only marketing but also seeking out potential licensees, including overseas (*e.g.* MIT in the United States, the Fraunhaufer in Germany).

- The third approach consists in the creation of *public or private intermediaries* to support technology transfer and licensing. Examples of private intermediaries include the Research Corporation Technologies and, more recently, Internet-based platforms in the United States. Examples of public intermediaries include the Austrian patent exploitation agency, Tecma, which provides support for the assessment of potential inventions from universities. Austrian university researchers are not, however, required to disclose their inventions to Tecma. In Japan, the government has sponsored the creation of large network of Technology Licensing Offices (TLOs) to encourage and formalise technology transfer between universities and industry. In some cases, these transfer institutions have a mandate to support SMEs, as is the case for the Intellectual Property Services Office in Canada.

While the aim of public licensing agencies is to fill the gap where there is insufficient critical mass within universities to support such activities, developing expertise and a sufficient customer base to generate revenue will require sustained levels of investment, mainly from public support. Another issue relates to their distance from research institutions as this can limit their role in educating researchers about potential commercial applications. In addition, these agencies may experience difficulties in competing with private sector intermediaries, not only in terms of seeking clients but also for hiring the technical skills they need, especially technology examiners trained in rapidly changing fields.

NOTES

1. "Knowledge spillovers" refer to a number of processes and infrastructures that have in common that they facilitate "informal knowledge transactions" between industry and the science system: science parks; incubators; firms' laboratories on campus; public laboratories as lead users of innovative equipment; informal interactions between public research staff and industry's researchers.

2. Reynolds, P. *et al.* (1999), *Global Entrepreneurship Monitor, 1999 Executive Report*, Kauffman Center for Entrepreneurial Leadership, 1999.

3. The disincentive to disclosure and protection inherent in the Japanese system of individual-based ownership of IPRs is often cited as being the main reason for low patenting by universities but also for the importance of informal channels through which professors grant their IPRs to companies in exchange for donations to their research departments (Kneller, 1999).

BIBLIOGRAPHY

ADL (Arthur D. Little International, Inc.) (2000), "Good Practice in Technology Transfer from Large Public Research Institutions (LPRIs)", unpublished report for the European Commission.

Austrian Federal Ministry of Economy and Labour (2001), *Benchmarking Industry-Science Relations - The Role of Framework Conditions*, Research Project commissioned by the European Commission, Enterprise DG and the Federal Ministry of Economy and Labour, Austria; project managed by the EC Benchmarking Co-Ordination Office (http://www.benchmarking-in-europe.com) and co-ordinated by the Institute of Technology and Regional Policy Joanneum Research, in co-operation with the Centre for European Economic Research (ZEW) and the Austrian Research Center Seibersdorf (ARCS), forthcoming.

Danish Ministry of Research and Technology (2000), *University Performance Contracts: The Danish Model*, Statens Publikationer, Copenhagen.

David, P. and D. Foray (1995), "Accessing and Expanding the Science and Technology Knowledge Base", *STI Review*, No. 16, Special Issue on Innovation and Standards, OECD, Paris, pp. 13-68.

GAO (US General Accounting Office) (1999), "Technology Transfer: Reporting Requirements for Federally Sponsored Invention", August, http://www.gao.gov/audit.htm

Kneller, Robert (1999), "Intellectual Property Rights and Incentives for University-Industry Technology Transfer in Japan", in L. Branscomb, F. Kodama and R. Florida (eds.), *Industrializing Knowledge: University-Industry Linkages in Japan and the United States*, The MIT Press, Cambridge, Mass.

Mustar, P. (2000), "Rapporteur's Report on the Workshop on Spin-offs", German/OECD Conference on Benchmarking Industry-Science Relationships, Berlin, October.

Observatoire des Sciences et Techniques (2000), *Indicateurs 2000*, Economica, Paris.

OECD (2001a), *Innovative Clusters: Drivers of National Innovation Systems*, OECD, Paris.

OECD (2001b), *Innovative Networks: Co-operation in National Innovation Systems*, OECD, Paris.

OECD (2001c), *Innovative People: Mobilising Skills in National Innovation Systems*, OECD, Paris.

OECD (2002), *International Mobility of the Highly Skilled*, OECD, Paris.

Polt, W., C. Ramer, H. Gassler, A. Schibany and D. Schartinger (2001), "Benchmarking Industry-Science Relations: The Role of Framework Conditions", in Special Issue on the Benchmarking of RTD Policies in Europe, *Science and Public Policy*, Vol. 28, No. 4.

Porter, M. and S. Stern (1999), "The New Challenge to America's Prosperity: Findings from the Innovation Index", US Council on Competitiveness.

Rammer, C. (2001), *Industry-Science Relations in Germany – An Overview*, Centre for European Economic Research (ZEW), Mannheim, June.

SPRU (2000), "Talent, Not Technology: Publicly Funded Research and Innovation in the United Kingdom", Science and Technology Policy Research (SPRU), University of Sussex.

Teaching Company Scheme (2001), *TCS Annual Report 2000/2001*, UK Department of Trade and Industry, October.

Thorburn, L. (1999), "Institutional Structures and Arrangements at Public Sector Laboratories", paper presented at the TIP Workshop on High-technology Spin-offs from Public Sector Research, Paris, 8 December.

Chapter 3

PILOT STUDY ON FRANCE AND THE UNITED KINGDOM[*]

As part of the OECD project on benchmarking industry-science relationships (ISRs), the objective of this pilot study was to: *i)* experiment with benchmarking as a tool for policy diagnosis; and *ii)* contribute to the development of a benchmarking methodology suitable for public policy purposes. The benchmarking process (Figure 1) involved a series of seminars,[1] fed by analytical and methodological papers, including two country reports (see Part II). The following highlights the results of the pilot study regarding both benchmarking methodology and policy diagnosis.

Figure 1. **The benchmarking process**

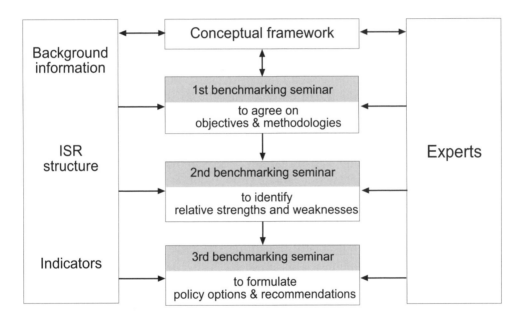

———————————

*   This chapter has been written by Rémi Barré and Jean Guinet. The views expressed are not necessarily those of the other participants in the benchmarking exercise.

Benchmarking as a joint learning and strategic evaluation process

Benchmarking is not a "beauty contest", based on crude indicators, but rather a learning process through which a public or private organisation can compare itself to its counterparts with a view to finding novel ways and motivations to improve its performance. When applied to government innovation policy, benchmarking is expected to yield several benefits:

- Improved understanding of the strengths and weaknesses of a national innovation system.

- Identification of good practices and the institutional changes needed to implement them.

- Guidelines for the development of new indicators.

- Extended international networks of innovation policy makers and analysts.

Meaningful benchmarking must avoid two traps. It is important not to view benchmarking as a simple arithmetic calculation, founded on a few basic indicators. This approach assumes that ISRs can be characterised by a few simple input and output indicators, and ranked according to measured performance. It has the advantage of producing results that are easy to communicate to decision makers; however, its cognitive content is extremely weak and it can lead to inappropriate policy recommendations. In addition, it does not create incentives for improvement since the actors do not recognise themselves in such a simplistic procedure.

The second trap involves looking for depth at the expense of international comparability, and consequently being overwhelmed by country idiosyncrasies. In this approach, benchmarking is reduced to a set of case studies which can only be put side by side. While such an approach can mobilise actors and provide interesting information, it does not challenge current practices and thus creates little "peer pressure" on policy makers.

The challenge for the pilot study was to find a way around these traps. This was done through a structured and iterative process of interaction among stakeholders from the two countries in addressing common issues, based on both indicators (quantification – codified knowledge) and qualitative information and expert judgement (contextualisation – tacit knowledge). It has been decided to concentrate on three key issues: *i)* ISRs and education; *ii)* responsiveness of public research to industry and social needs; and *iii)* instruments for carrying out industry-science transactions.

Strengths and weaknesses of the French and UK ISR systems

ISRs and education: the fabric of industry's top management

France

A dual higher education system [universities and *Grandes Écoles* (GEs)] has shaped French ISRs, with enduring consequences. In contrast to the university system (with the exception of medicine and political science), students enter the GEs after highly competitive examinations. Traditionally, GEs have strong ties with industry and most of their graduates go on to obtain high-level positions in industry. Until recently, the GEs have had relatively little research capacity while the reverse is true of the universities. Since PhDs are produced by the universities, which have weaker ties with industry, only a minority of PhDs end up working in firms. The best PhDs target a career in the public sector, and the internships offered to university students to spend a year in a firm have only a modest uptake, with some internships not being taken up.

The linkages between the GEs and industry are most effective in relatively mature technological areas, such as aeronautics, automobile, chemistry, electrical machinery, professional electronics, defence, agro-food. The higher education system provides the young talent and new knowledge which is essential for technological innovation in these industries.

This is not the case in the emerging industries. The new sectors of the economy (some fields in biotechnology, software and several sub-sets of the service sector, etc.), mainly have to rely on the universities, which are usually far less experienced in handling relations with industry, both in terms of education and knowledge transfer for innovation.

Although a number of common PhD programmes have been set up between the GEs and the universities, there is a clear need to strengthen the linkages between the GEs and academic research through contacts with the universities or with the CNRS (*Centre National de la Recherche Scientifique*), and improve the linkages between universities and industry, either directly or through partnerships with the GEs.

United Kingdom

Universities are the "standard" higher education institution, and the PhD the "standard" higher diploma. This highlights the issue of the educational profile of top managers in UK-owned firms, who have rarely been trained as scientists, in contrast with US executives, who have often both a PhD and a Master of business administration (MBA), or with their French counterparts. A major problem is that firms tend to have an insufficient capability to absorb science, reflecting the rather low level of R&D carried out in most companies. In fact, as is the case in many other countries, foreign subsidiaries show a greater propensity than domestic firms to collaborate with universities.

On the other hand, since UK universities traditionally maintain close contacts with industry – where the majority of their students end up working – and since employment of postdocs and interns is a flexible process, it is easy to understand why emerging science-based industries like the biotechnology and information and communication technology (ICT) sectors see the universities as a favourable pool human resources. Hence, in these sectors UK industry is particularly innovative and competitive, while the more traditional activities suffer from a lack of technological vision and engineering skills. Contrary to the French situation, the problem seems to be with the existing, mature industries.

One of the explicit aims of the UK foresight exercise is to bridge the gap between the research community and senior management in firms, seen as a barrier to better ISRs. The limiting factor to ISR is considered to be the will and the capabilities of firms (with the exception of the science-based sectors).

That being said, it would appear that the incentive systems applied in the universities, and in particular the research assessment exercise, with its focus on academic excellence in the disciplinary tradition, do not encourage the development of synergies between education, academic research and industry.

Table 1. **Comparative ISR indicators**

	France	United Kingdom
Total researchers per thousand labour force (1999)	6.14	5.54
Total new S&T PhDs per thousand population aged to 34 years (1999)	0.71	0.63
R&D intensity (%) (1999)	2.17	1.87
R&D expenditure – average annual real growth (%) (1995-99)	0.62	1.23
Industry-financed R&D as a percentage of industrial output (1999)	1.57	1.27
Industry-financed R&D – average annual real growth (%) (1995-99)	3.48	1.99
Industrial RD financed by public contracts (%)	7.7	2.1
Participation to the European RTD programmes (%)	15.3	16.4
SMEs in publicly funded R&D executed by the business sector (%) (1995-99)	8.33	10.2
European patents per million population (1999)	118	95
World share of European patents invented in key technologies	6.5	5.8
Average annual growth (%) in European patents (1995-99)	7.59	11.16
Number of scientific publications per million population (1999)	652	949
Average annual growth (%) of number of scientific publications (1995-99)	2.74	1.52
Highly cited papers as a percentage of total number of publications (1997-99)	1.1	1.5
Number of highly cited publications per million population (1999)	26	54
Percentage of innovating firms co-operating with other firms, universities or public research institutes (1996)	30	31
Knowledge-intensive services employment in total employment (%) (1995-99)	34.7	39.3
World market share of exports of high-tech products (%) (1999)	7.39	6.31

Source: European Commission, *Towards a European Research Area, Key Figures 2001 – Indicators for Benchmarking of National Research Policies*; OST, *S&T indicateurs 2000.*

Other features of the UK system tend to pull in the opposite direction. These include: possibilities for students to easily interrupt their studies; possibilities for students to take up an internship, even prior to the completion of their diploma; and the number of part-time, non-permanent professor positions. In the same spirit, the CASE (Collaborative Awards in Science and Engineering) and TCS (Teaching Company Scheme) are key policy tools, along with the Link and Faraday programmes and company initiatives. The primary principle of these schemes is competitive bidding, at either the individual or the institutional level. There is some concern that this trend could accentuate the gap between those that participate in ISRs and those that do not, at the expense of a wider diffusion of knowledge.

Responsiveness to industry and social need: governance of the public research system

France

Beyond the often-highlighted centralisation syndrome, the real picture of the governance of the French research system is in fact rather blurred. One important feature of the French system is the importance of the specialised public research organisations (PROs) that enjoy a wide autonomy in decision making regarding research priorities. The PROs have a clearly defined sectoral mandate (transport, agriculture, health, maritime issues, energy, building technology, ICTs, etc.).

Two main questions arise: Do the PROs respond adequately to evolving social needs, and in particular to the needs of industry? Have they adapted to the emergence of new science-based sectors? These are challenging issues as there may be a partial contradiction between the organisational principles of some of the PROs (including the quest for scientific excellence) and their mandate, and because the PROs are members of longstanding innovation networks, including the *Grandes écoles d'ingénieurs* (GEIs), that serve the traditional industries.

R&D contract funding by the ministries, the regions, the European Commission and industry can significantly influence the orientation of public research, especially at laboratory level, since firms generally pay only the marginal costs of research. On the other hand, the responsiveness of the public research sector is limited by the fact that researchers are civil servants and elect two-thirds of the members of scientific committees in charge of prioritisation. Again, the picture is a complex one, and different rationales are operating at one and the same time.

Research teams and laboratories created under joint public–private initiatives, as well as firms set up as subsidiaries of PROs, constitute valuable instruments for co-operative research. Another important mechanism, *i.e.* funding by the regions through the planning contracts between the regions and the State (*Contrat de Plan État-Régions*), has helped the universities and PROs to better understand socio-economic context. These contracts play an important role in higher education, research and technology transfer. Generally speaking, the French system has a strong regional dimension. Similarly, the four-year contracts (*contrats quadriennaux*) agreed between the research institutions (universities and PROs) and the government provide a framework for discussion and negotiation that take into account the socio-political environment of research activities. However, it should be noted that the focus on *ex-ante* evaluation of these contracts and insufficient monitoring and *ex-post* evaluation diminish their efficiency with regard to the fulfilment of broad socio-economic objectives.

Most academic laboratories are "mixed", *i.e.* they obtain resources and personnel from both the university and the PROs (mainly the CNRS). This situation affects the governance of the labs and can either improve or diminish their responsiveness, depending on the institutional configuration. Two other features of French university research may limit its responsiveness to the needs of industry:

- In academic research, there is often a perceived trade-off between "meeting the needs" of industry and pursuing high-quality "prestige" research, all the more so since mainly elected researchers are involved in official scientific evaluation and orientation of programmes.
- University presidents (chancellors) have only limited powers to set up and implement a strategy. The 1999 Law on innovation contains provisions designed to ease the handling of industry–science relationships (ISRs) by universities (through the so-called SAIC – S*ervices d'activités industrielles et commerciales*), but the "decree of application", required for the implementation of this part of the Law, has yet to be published.

United Kingdom

The governance of the publicly funded research system is less decentralised than one might think, due to the role of the research councils and of the research assessment exercise.

The universities and Research Councils respond to signals from the government, when it uses clear criteria for allocating public money to them. In fact, these institutions balance the various signals, including those coming from the research community. Since industrial funding is the only category of funds which, until recently, could increase significantly, universities have tended to become more receptive to industrial needs. In the same line, the "teaching quality reviews" of the universities include criteria about the jobs held by students. This provides a clear incentive for universities to enhance their industrial relations.

However, many of the non-financial elements of the incentive structure at all levels (institution, laboratory/department, individual researcher) pull in the opposite direction, in particular the Research Assessment Exercise (RAE). Furthermore, firms need inter-disciplinarity, but the majority of the universities are ill-equipped to respond to this demand.

This is where the start-ups and the knowledge-intensive business services (KIBS) have a competitive edge. This is also why spin-offs are important, since they create a direct link between research and markets. However, they require public support during their infancy, and the question arises as to the appropriate selection process until such time as the market can play its role.

The universities tend to specialise in certain types of industry-science relationships (ISRs), in order to find a niche in the competition for funds.

An important evolution is taking place regarding the incentives for ISRs, which may be changing the institutional context: the Higher Education Innovation Fund (HEIF) is a platform for core funding designed to enable the universities to pursue ISRs in favourable conditions (the "third leg" of funding, after core and contract funding). This is important because it provides the ISRs, not only with greater funding but also with a stronger political and institutional status.

Another evolution which is serving to improve the universities' grasp of socio-economic needs is the growing importance of regional government support in innovation and thus, of regional constituencies. This trend will give greater weight to local industrial needs.

Instruments for carrying out industry-science transactions

France

Today, initiatives are burgeoning in most of the higher education and research institutions: the development of incubators at the regional level and seed funding has been rapid in the last two years; tax incentives have also been instrumental in promoting venture capital; the numbers of contractual arrangements with industry, but also other partners, have increased. Starting from a set of general guidelines and a few specific tools, the local actors have developed a variety of practices.

Consequently, the boundaries between "public" and "private" are becoming less clear for researchers and research managers. Laboratories are finding it difficult to handle the new and often complex contractual arrangements. At the same time, the pressure to hire contract personnel is growing.

The lack of flexibility provided by the employment regulations is becoming a serious problem, since there are few options between tenure (including for technicians and administrative personnel) and very restrictive (non-renewable) fixed-term contracts. Hence, many institutions work – officially or not – with "associations", *i.e.* non-profit organisations which manage ISR contracts and can hire personnel on standard private employment contracts. There is a need to better monitor and assess these new trends in order to learn from the various experiences.

Finally, the following issues also deserve consideration:

- The development of incubators and seed funding has been rapid in the last two years. Public/private partnership and decentralisation are key to success in these areas, and it is important to broaden industrial participation beyond the "traditional circles".

- Tax incentives have been instrumental in promoting venture capital. However, the question of the tax treatment of private investment in incubators has been neglected.

- The four-year contracts (*contrats quadriennaux*) between the Ministry of Research, a university and a public lab (usually the CNRS) state which institution is responsible for handling contracts for external funding. This simplifies governance but, in some cases, can come at the expense of flexibility.

United Kingdom

Private subsidiaries of universities are becoming central actors in industry–science relationships, although the consequences of this trend remain uncertain. Nevertheless, a key feature of the handling of ISRs at the local level is its diversity: in some universities, the management of ISRs is left to the individuals concerned; in others, all aspects of ISRs are centralised in a specialised office or in a subsidiary; in yet others, an intermediate situation exists, mixing aspects of the above. It is noteworthy that a number of universities have hired industrialists as vice-chancellors, with a view to facilitating the creation and development of ISRs.

In some cases, the limits to the development of ISRs may have been reached; not all universities should be pushed to develop instruments for the commercialisation of research. It is important to recognise that universities have other missions such as teaching and research, which are just as important as developing relationships with industry.

The question of whether government incentives should be directed primarily to those who have already begun a successful venture, or to those that need to be encouraged to undertake a project, is not being specifically addressed.

In response to the interest shown at the highest level of government, there has been a profusion of schemes to promote spin-offs, start-ups, incubators, venture capital, science parks. These have not been properly evaluated, but there is no evidence of substantial economic impact as yet.

Finally, it must be said that, as the US experience shows, intellectual property rights (IPR) are potentially a problem for ISRs.

Synthesis of results and public policy implications

The formal structures of higher education and research exaggerate the differences between the two countries in terms of IRs: the legal and institutional set-ups are quite different but the political and sociological contexts are similar. Thus, the policies have similar aims, although their specific forms vary to accommodate the differing structures.

Shared diagnosis and overall vision

ISRs have been a concern for at least the last 20 years in both countries. The role of public research in enhancing industrial innovation capabilities has been long recognised as a crucial area for public policy.

ISRs and education: the fabric of industry's top management

ISRs are shaped by the educational path which leads to top managerial positions in industry: the role of science in education and the linkages between industry and the higher education institutions are key to securing solid, long-term relations between firms and external sources of technological innovation. During the years spent in education, informal innovation networks are built up which last throughout one's professional life, linking industry to technology and the research base.

Innovative firms want to improve their access to the best students and researchers. However, there is a danger that this could lead to a brain drain from academia and to further stratification in the education and public research system. This makes the question of how to reconcile excellency and broad-based upgrading of education levels and innovative capabilities are the more urgent.

Responsiveness to industry and social needs: governance of the public research system

In industry-science relations, the optimum is not the maximum. The all-important role of higher education institutions is to provide appropriate education and training. Too systematic and over-riding an orientation towards industry relations could jeopardise this primary role. In addition, the universities should not respond to industry's current needs at the price of weakening their capacity to address future needs and challenges.

It is recognised that it is necessary to avoid generalisation and that three types of ISRs can be distinguished:

- *Relations between multinational enterprises (MNEs) and world-class universities.* MNEs are externalising part of their research and development activities and are looking for the best laboratories, scientists and students. Their concern is not whether specific universities "meet their needs", but that they are world-class.

- *Relations between universities and small, high–technology firms (spin-offs and knowledge intensive business services – KIBS).* This phenomenon – which is important in terms of quality, although small in terms of size – will have to be assessed in a longer-term perspective.

- *Relations developing in a regional context between firms (often SMEs) and the local university.* Here, firms are looking for shorter-term, problem-solving capabilities. In other words, the relationship is one based on needs.

ISRs entail different rationales and different mechanisms depending on the technological fields and innovative clusters involved. This points to the need for a fairly decentralised governance of ISRs, with different organisational models and different instruments.

Instruments for carrying out industry-science transactions

Government-induced and spontaneous developments over a long period and more than two decades of policies aimed at developing ISRs have resulted in a similar situation in both countries: ISRs are now part of the "academic landscape", but significant difficulties have yet to be solved.

ISRs are an integral part of the academic landscape:

- Income from industry has become important for public research organisations and higher education. Contractual income has become part of the financial structure of universities and public research and is necessary to balance their accounts.

- There has been a "culture change" within academia and ISRs have become one of the activities considered by academic staff as part of their mission.

Key difficulties requiring solution include:

- Researchers, academic staff and universities themselves are often confronted with contradictory incentives regarding ISRs.

- There are concerns about the costs of exploiting and protecting industrial property rights (IPRs) and about possible limitations to the diffusion of knowledge which could be harmful to further academic research.

- The proliferation of intermediary structures for implementing ISRs could inhibit or distort, instead of stimulating, ISRs.

- Indicators needed to monitor trends in ISRs and evaluate policy initiatives and instruments are on the whole insufficient. Comparable indicators for the various universities are lacking.

- Adequate accounting systems in universities are lacking, impeding strategic management of ISRs at the local level and transparency *vis-à-vis* public policy makers.

Finally, one shared view is the need for initiative at the university or institution level, considering that the disciplinary and industrial structure specificities entail specific ISR practices, which can be best decided and implemented at local level.

Policy principles and instruments: a comparison

Broadly speaking, ISR policy in the two countries is based on the same concepts, addresses the same problem areas and has the same structure (Table 2). However, there are three main differences at the policy instrument level.

The first concerns the combination of actions within a given instrument. For example, the UK Teaching Company Scheme (TCS) has both a networking and a human resource component, whereas the French Network for Research and Technological Innovation (RRIT) scheme is solely a network-building instrument, a separate instrument (CIFRE) being devoted to the financing of doctoral students working in industrial laboratories.

A second major difference relates to the scope of a given instrument. A striking example is technology foresight. In France, the term refers to studies of specific technological fields (carried out by working groups comprised of a small number of experts); in the United Kingdom, it refers to an ambitious national undertaking involving thousands of people over two periods of three years.

Table 2. **ISR policy objectives and instruments in France and the United Kingdom**

Policy objectives	Instruments	
	France	United Kingdom
Building lasting linkages between public-private research; networking	• Networks for research and technological innovation (*réseaux de recherche pour l'innovation technologique* – RRIT) • Common public-private laboratories scheme (*laboratoires communs*) • National centres for technological research located in regions (CNRT)	• LINK programme • Teaching Company Scheme (TCS)
Fostering industry-science relations through individual, informal contacts	• Industrial relations club (*club des relations industrielles* – ECRIN) • National association for technical research (ANRT)	
Training young researchers for ISRs	• CIFRE scheme (university-industry doctoral programme) • COTRTECHS scheme (for technicians)	• Teaching Company Scheme (TCS) • Collaborative Awards in Science & Engineering (CASE)
Organisation of the provision of technology transfer by public research	• Industrial technical centres (*centres techniques industriels* – CTIs) • Technological diffusion networks (*réseaux de diffusion technologique* – RDT) • Regional technology transfer and innovation centres (CRITT)	• Faraday Partnerships • Manufacturing technology centres (under discussion)
Organisation and enhancement of the universities' capability to manage ISRs	• Intellectual property rights (IPR) management subsidiaries • University-industry relations bureau (*services d'activités industrielles et commerciales* – SAIC)	• Science-enterprise challenge (SEC) • Higher Education Innovation Fund (HEIF) ("third leg" of financing) • Higher Education Reach Out to Business and the Community (HEIF)
Fostering industrial innovation through the development of spin-out firms	• Capital for early-stage development (*fonds d'amorçage*) • Funding of public incubators (*incubateurs issus de la recherche publique*)	• University Challenge Fund
Foresight activities	• Delphi exercise (1996) • Key technologies studies (1995 and 2000)	• Technology foresight (1995-98) • Foresight (1999-2001)

The third difference lies in instrument design. Two instruments addressing the same problem area may be quite different in design because they have to operate in a different institutional context, especially in terms of the financing and steering of public research organisations (*e.g.* the dependence on competitive bidding in the United Kingdom). Financial incentives are more likely to be used in the United Kingdom than in France, where regulatory changes or conventional subsidies are the preferred approach – although there are some exceptions (*e.g.* calls for tender and competitive bidding have been used to finance incubators). In the United Kingdom, the tendency is to state the objectives to be reached, leaving the universities free to propose ways to achieve them, while in France, the means required to reach the objectives are generally stated. In France, regional initiatives play an important role, albeit one that is not always easy to monitor (the existence of centralised mechanisms should not serve to hide the many *ad hoc*, locally defined, instruments and procedures). This "regionalisation" of ISR policy seems to be in its infancy in the United Kingdom.[2]

Concluding remarks

Developing a benchmarking culture

The ISR policy instruments together constitute a system; the efficiency of a particular instrument needs to be considered within this broader framework. This comparison between ISRs in France and the United Kingdom has shown that different combinations of actions can be used to build integrated policy instruments. In this context, it is important to promote a benchmarking culture among all stakeholders. Governments should organise and encourage benchmarking at the institutional – or micro – level (as stressed at the Berlin Conference). More frequent exchanges of experience between managers of the same category of instruments in the two countries should be complemented by regular benchmarking seminars where the performance of individual policy instruments could be compared and discussed from a systemic perspective.

Developing benchmarking indicators

The existing indicators in this area are particularly ill suited for meaningful international comparisons. In addition, they are not sufficient to inform a policy-oriented discussion; relevant information is often held by the research institutions and there is a need to collect and aggregate such information in order to draw clearer national pictures. Significant indicators should exploit data on patenting, licensing, spin-offs, co-publications between industry and university, citation of industry papers by academics, labour mobility and financial flows. Data from the Community Innovation Surveys (CIS) can also provide useful indicators of interactions between public research institutions and industry.

MAJOR PUBLIC PROMOTION ACTIVITIES FOR INDUSTRY-SCIENCE RELATIONSHIPS IN THE UNITED KINGDOM

Activity	Year introduced	Objectives	Type(s) of industry-science interaction
Foresight	1993	• To develop visions of the future – looking at possible future needs, opportunities and threats, and deciding what should be done now to make sure we are ready for these challenges • Build bridges between senior people in business, science and government, bringing together the knowledge and expertise of many people across all areas and activities	Networking (but limited involvement of commercial management)
LINK	1985	• To encourage research collaboration between industry and the science base	Collaborative research
Faraday Partnerships	1999	• Intermediary organisations hosting research and technology adoption and translation activities (which can involve universities)	Technology adaptation, facilitating collaborative research, personnel mobility, training & education
University Challenge Fund	1999	• Provides support to universities or consortia of universities to set up local seed funds supporting early-stage commercialisation	Knowledge transfer through spin-outs, IPR, develop prototypes

Science Enterprise Challenge (SEC)	1999	• To encourage transfer of S&T innovation to the business sector by establishing "centres of enterprise" in universities to:
		  ▪ teach enterprise and entrepreneurship to science and technology students
		  ▪ make ideas and know-how available to business to support competitiveness and wealth creation
		  ▪ encourage the growth of new businesses by supporting start-ups, including spin-out companies based on innovative ideas developed by students and faculty within the universities
			Training & education, technology transfer

Higher Education Innovation Fund (HEIF)	1998	• Funding for the establishment of centres of expertise in HEIs, ISR-oriented training for HEI staff, "one-stop shops" for business partners	Technology transfer, contract research, networking, personnel mobility
Joint Research Equipment Initiative (JREI)	1996	• Funding of equipment in areas of high-quality research	Contract research, collaborative research
Collaborative Awards in Science & Engineering (CASE)	n.a.	• Provides grants to students carrying out doctoral research addressing industrial problems and jointly supervised by HEIs and firms • There is also an Industrial CASE studentship scheme where industrial partners choose an academic partner for research training, and a CASE for New Academics scheme that provides a route for new academics to build links with a company at an early stage in their career through co-supervision of a CASE student	Training & education

MAJOR PUBLIC PROMOTION ACTIVITIES FOR INDUSTRY-SCIENCE RELATIONSHIPS IN FRANCE

Activity	Year introduced	Objectives	Type(s) of industry-science interaction
Industrial technical centres (*centres techniques industriels* – CTI)	1975	• To facilitate the sectoral transfer of technology on a national basis, and encourage the spread of technical and management skills • To enhance the quality of research and training in science, engineering and technology by stimulating collaborative research and development projects	Technology transfer
Collaborative research units (*laboratoires mixtes recherche publique/entreprises*)	1985	• To set up special research teams or units created under both private and public initiative (public research institutions)	Co-operative research
Foresight – Delphi	1995	• To forecast the emergence of new technologies	Collaboration between industry managers and public research managers
Foresight – Key technologies (*technologies-clés*)	2000	• To facilitate interactions between scientific, technological and market prospective visions (to 2005)	
Networks for technological diffusion (*réseaux de diffusion technologique* – RDT)	1990	• To encourage regional research collaboration between SMEs and technology advisors	Collaborative research

Networks for research and technological innovation (*réseaux de recherche et innovation technologique* – RRIT)	1998	• To set up networks of excellence gathering the main public research organisations, with a view to:
		  • defining priority measures
		  • developing new collaborative efforts between public research and firms, including SMEs
			Co-operation through networking

National centres for technological research (located in the regions)	1999	• To enhance collaboration between public laboratories and large/small high-technology firms carrying out world-class R&D	Collaborative research

Seed financing (*fonds d'amorçage*) (three national funds, seven regional funds)	1999	• To provide financial support as early-stage risk capital to SMEs, especially spin-offs from public research	Spin-offs, IPR, prototypes
Technology transfer and innovation regional centers (*centres régionaux de recherche et d'innovation technologique* – CRITT)	1) 1983 2) 1993	• To provide technological services to SMEs, in collaboration with CTI (see above), universities and institutes of technology, acting as intermediaries or service providers • (Some of the CRITT have obtained a special label, "*Centres de ressources technologiques*" – CRT)	Knowledge transfer
Public incubators (*incubateurs issus de la recherche publique*)	1999	• To encourage firm creation (spin-offs)	Training & education, technology transfer
Public research institution-enterprise relations bureau [FIST-CNRS, INRIA-Transfer, INRA(ATI), etc.]	From 1990	• To facilitate the commercialisation of results from research in public labs	Technology transfer, IPR, personnel mobility
University-enterprise relations bureau (*service d'activités industrielles et commerciales* – SAIC)	2000	• To facilitate the commercialisation of results from research in higher education	Technology transfer, IPR, personnel mobility
Specific associations for promoting ISRs: ANRT, ECRIN, ISF		• To facilitate dialogue between the private and public sectors	Co-operative activities
Conventions for industrial research in universities (CIFRE)	1985	• To provide grants to students carrying out doctoral research addressing industrial problems and jointly supervised by HEIs and firms.	Training, mobility of highly skilled labour
CORTECHS scheme (for technicians)		• To facilitate high-level industrial and technological training through research	

COMPARISON BETWEEN PUBLIC RESEARCH
IN FRANCE AND THE UNITED KINGDOM:
KEY FIGURES REGARDING THE SCIENTIFIC PERSONNEL

Table A3.1. **Scientific personnel in academic research, government laboratories and industry**

Percentages

	France	United Kingdom
Academic research*	43	48
Government labs	15	7
Industry	42	45
Total	100	100
Number (in thousands)	161.5	177.6

* Including CNRS in France.

Table A3.2. **Scientific personnel in academic research and government laboratories**

Percentages

	France	United Kingdom
Academic research*	75	87
Government labs	25	13
Total	100	100
Number (in thousands)	93.0	97.6

* Including CNRS in France.

Table A3.3. **Research only staff, teaching & research staff, academic research**

Percentages

	France	United Kingdom
Research only	19	38
Teaching & research	81	62
Total	100	100
Number (in thousands)	70.0	84.6

Table A3.4. **Academic research only staff, teaching & research staff, government laboratories**

Percentages

	France	United Kingdom
Research only staff	14	33
Teaching & research	61	54
Government labs	25	13
Total	100	100
Number (in thousands)	93.0	97.6

Table A3.5. **Contract *versus* permanent academic staff**

Percentages

	France	United Kingdom
Contract staff	17	45
Permanent staff	83	55
Total	100	100
Number (in thousands)	70.0	84.6

NOTES

1.   Participants at the series of seminars included:

France: R. Barré (OST), M. Barret (INRIA), M. Carisey (CNRS), F. Cherbonnier (Ministry of Finance), D. Coulomb (Deputy-Director, Ministry of Research and Education), P.Y. Mauguen (Ministry of Research and Education), J.C. Pomerol (Université Paris VI).

United Kingdom: J. Barber (Director, DTI, Chairman of the OECD Committee for Scientific and Technological Policy), C. Bryant (DTI), D. Clarke (Rolls-Royce), R. Freeman (Intellectual Property Right Institute), L. Georghiou (PREST, University of Manchester), F. Ismael (DTI), P. McCoy (British Aerospace), K. Pavitt (SPRU).

OECD Secretariat: Jean Guinet.

2.   With the exception of Scotland, Wales and Northern Ireland.

Part II

BENCHMARKING INDUSTRY-SCIENCE RELATIONSHIPS

COUNTRY CASE STUDIES

Chapter 4

INDUSTRY-SCIENCE RELATIONSHIPS IN FRANCE*

1.   General introduction

For the last few years, issues regarding the relationships between industry and science have been high on the policy agenda in France, reflecting a growing awareness of the importance of the linkages between public research and industry as a determinant of the innovation capacity and competitiveness of the economy. This has led to a number of initiatives at the national and regional levels by government and public research organisations to improve the efficiency of ISRs. These initiatives were inspired by the experience of other countries, particularly the Anglo-Saxon ones, which are recognised as being the champions of ISRs. However, the specificities of the French institutional set-up do not facilitate the transfer of experiences from abroad; many of the lessons learned through comparative assessment and missions abroad need to be "translated" before they can be applied. This is being done through a learning process that also involves comparisons among French regions and research organisations. However, this approach is impaired by the lack of a conceptual framework and indicators that would provide all stakeholders with a common basis for their own benchmarking efforts. The OECD pilot study provided the opportunity to develop such a framework in a way that facilitates international comparisons.

This chapter provides some basic information on ISRs in France by: describing the knowledge-producing institutions and their main channels of interaction (Sections 2 to 4); characterising the ISR incentive and organisational structures (Section 5); and illustrating the role of ISRs in training human resources and that of labour mobility in enhancing ISRs (Section 6). The chapter has benefited from comments by the working group on *indicateurs de valorisation*, in which the majority of French public research laboratories, universities and the Ministry of Research are represented.

2.   Knowledge production in France

2.1.   *Institutional and funding structures*

Research funding

National R&D spending (GERD) amounted to EUR 29 billion in 1999, corresponding to 2.2% of GDP, falling from 2.4% in 1990 (it was 2% in 1980). The decrease in the GERD/GDP ratio during the 1990s was mainly due to the sharp decline in defence R&D spending. In fact, the share of GERD financed by firms increased from 45% to 51% in the 1990s. For comparison, the GERD/GDP ratio for France is below that of the United States, Japan, Sweden, Finland, Switzerland and Germany. The

---

*   This chapter has been written by Rémi Barré of the Observatoire des Sciences et des Techniques, France. Any errors in figures, facts or interpretation are those of OST and the author.

share of GERD that is financed by firms remains significantly below that of the most developed countries (60-70%, compared with around 50% in France). The major components of GERD in France are (Table 1):

- Research with academic objectives [research in the higher education institutions and the *Centre National de la Recherche Scientifique* (CNRS)], amounting to 17% of GERD.

- R&D in support to public policy in priority areas (environment, health, agriculture, etc.), amounting to 11% of GERD.

- R&D aimed at fulfilling national strategic objectives through "large technology programmes" (LTPs – *grands programmes technologiques*) of a civilian or military nature. This consists in the development of complex technological systems for which the state is the main – and sometimes exclusive – client. The strategic objectives relate to defence, as well as to technological and economic sovereignty. The related R&D spending, all of which is publicly financed, represents 22% of GERD, 40% of which is subcontracted to firms through large R&D contracts.

- R&D funded by firms for their own innovation objectives, amounting to 51% of the total.

Table 1. **Objectives of GERD, 1998**

Objectives of R&D funding	Volume (in EUR billion)	Volume (as a % of GERD)
Academic research	4.9	17%
Research for public policies	3.2	11%
R&D for the strategic objectives of the state (large technology programmes)	6.4	22%
Of which: defence financing	2.6	12%
Sub-total public financing	14.5	49%
R&D for the objectives of industry (funded by firms)	14.5	51%
Total	29	100%

GERD = gross expenditure on R&D.

The LTPs represent almost half of public funding for R&D (EUR 6.3 billion of a total EUR 14.3 billion). These programmes concern space, advanced aeronautics and nuclear activities, as well as specific information technologies, in addition to some other military R&D programmes (Table 2). With the exception of the last category, all the LTPs have both civilian and a military R&D funding. The space programme is the largest in terms of expenditures, representing 8% of GERD.

Table 2. **Large technology programmes (LTPs), 1998**

Large technology programmes (civilian and military budgets)	Volume (in EUR billion)	Volume (as a % of GERD)	Volume (as a % of public R&D spending)
Space	2.3	8%	16%
Aeronautics	1.2	4%	8%
ICTs/Telecommunications/Electronics	0.9	3%	6%
Nuclear R&D	1.2	4%	8%
Other defence R&D programmes	0.8	3%	6%
Total	6.4	22%	44%

Research execution

R&D is performed in four types of institutions: academic, targeted[1] public, military and industrial laboratories (Table 3).

Table 3. **R&D-performing institutions and laboratories**

Institution type	Laboratory type
Academic	Academic
Public research organisation (PRO)	Targeted public
Military institution	Military
Firm	Industrial

The first three types of laboratory, *i.e.* public research, account for 40% of GERD. This share has remained stable since the early 1980s; it is significantly higher than in comparable countries, in which public research accounts only for 20-35% of GERD. Furthermore, the volume of research performed in PROs (targeted public laboratories) is higher than that performed in academic institutions.

Table 4. **Gross expenditure on R&D by performing institution, 1998**

Institution/Laboratory	Volume (in EUR billion)	Volume (as a % of GERD)
Academic	4.9	17%
Public research organisations (PROs) – targeted public research	5.2	18%
Military	1.5	5%
Sub-total public research	11.6	40%
Firms – industrial laboratories	17.4	60%
Total	29.0	100%

Another way to characterise knowledge production activities is to measure the amount and allocation of human capital involved. It should be noted that, in what follows, only staff permanently engaged in public research are included (*i.e.* doctoral, postdoctoral students and contract researchers are excluded).

The picture is quite different from that resulting from the analysis of spending: 38% of scientists work in academic laboratories and 13% in PROs (Table 5). The figure for academic research is higher when doctoral students are included (there are few postdoctoral students and contract researchers).

Table 5. **Research personnel employed, by type of institution/laboratory, 1998**

Institution/Laboratory	Volume (number of scientists)	Number of scientists (%)
Academic	56 000	38%
Public research organisations (PROs) – targeted public research	19 000	13%
Military	3 000	2%
Firms – industrial laboratories	69 000	47%
Total (national)	147 000	100%

Public financing of R&D amounts to EUR 14.5 billion, and R&D performed by the public sector to EUR 11.6 billion, giving a net flow of EUR 2.9 billion of R&D contracts from the state to firms. This results from a gross flow of EUR 3.2 billion from the state to firms and EUR 0.3 billion from firms to public research institutions (Table 6). Public funding of industrial R&D (EUR 3.2 billion) is comprised of EUR 2.7 billion of R&D procurement by the large technology development programmes to private sector R&D subcontractors, and EUR 0.5 billion in research contracts provided to firms as an incentive to carry out research (incentive funding).

The EUR 11.6 billion of R&D conducted by the public research sector can be broken down as follows: EUR 10.2 billion of budgetary (core) funding, EUR 0.9 billion of research contracts from public sources (regional, national, European research programmes), and EUR 0.5 billion from industry.

Table 6. **Major financial flows in the R&D system, 1998**

EUR billion

Financing of R&D	Total funding	Performance of R&D by:	
		Public research institutions	Firms
Budget (core funding) of public research institutions	10.2	10.2	--
National incentive funds	0.4	0.2	0.2
Regional incentive funds	0.4	0.4	--
European incentive funds	0.5	0.3	0.2
Contracts from large technology programmes	2.8	--	2.8
Sub-total public financing	14.5	11.3	3.2
R&D funded by firms	14.5	0.5	14.0
Sub-total R&D funded by firms	14.5	0.3	14.0
Total financing	29.0	11.8	17.2

2.2.   *Public research in the higher education sector*

Academic research is performed by both the universities and the CNRS. University and CNRS staff collaborate closely, since 90% of CNRS personnel are employed in laboratories located in the universities.

In academic laboratories, 44 000 faculty members are employed as *maitres de conférences* (junior professors) and professors, engaged in both teaching and research. The CNRS employs 12 000 full-time researchers in all fields of science (except medical sciences) (Table 7). Academic research is undertaken in some 1 500 joint CNRS-university laboratories and in a number of purely university-staffed laboratories, employing approximately 40 000 doctoral students.

Evaluation of CNRS researchers (in terms of recruitment and career development) and of the joint laboratories is performed by the National Committee of Scientific Research, comprising some 40 specialised committees, of which two-thirds of the members are elected on a national level by the entire body of academic scientists (CNRS and faculty). Evaluation of faculty is carried out by the National Committee of Universities, consisting of about 60 specialised committees of which two-thirds of the members are elected on a national basis by faculty personnel only.

Table 7. **Scientific personnel performing academic research, 1999**

	Number of persons (in thousands)			Number of full time research equivalents[1] (in thousands)		
	Faculty (*enseignants-chercheurs*)	CNRS researchers	Doctoral students (of which 3 000 postdocs)	Faculty (*enseignants-chercheurs*)	CNRS researchers	Doctoral students (of which 3 000 postdocs)
Exact and natural sciences	22.0	9.0	19.5	11.0	9.0	19.5
Engineering sciences	7.0	1.0	7.0	3.5	1.0	7.0
Social sciences and humanities	15.0	2.0	13.5	7.5	2.0	13.5
Total	44.0	12.0	40.0	22.0	12.0	40.0

1. Each faculty member counts as 0.5 persons in full-time research equivalent.

Few postdoctoral students undertake academic research in France since contract research positions do not generally exist (with the exception of some positions offered for foreign visiting scientists, foreign young scientists for one year, and for some young French PhDs in the year following the completion of their doctorate).

There are about 80 universities and 150 engineering and agronomic schools (*Écoles d'ingénieurs* and *Écoles agronomiques*) in France. Three times as many engineering diplomas are delivered compared with PhDs in the exact and natural sciences (Table 8). In 1998, some 11 000 PhDs were produced in French universities, an increase of 30% since the early 1990s.

Table 8. **Engineering diplomas, graduates and PhDs, 1998**

	Social sciences and humanities	Exact, natural and engineering sciences	Total
Doctorates	3 700	7 400	11 100
Engineering diplomas	--	22 800	22 800
Graduates (diplôme de 3ème cycle: DEA and DESS)	33 900	17 100	51 000
Total	37 600	47 300	84 900

Among the 11 000 PhDs of the 1998 cohort, about one-third studied social sciences and humanities. More than a quarter were awarded to foreigners (an increase from one-third in the early 1990s). The proportion of foreigners is particularly high in the social sciences and the humanities, and in mathematics and engineering sciences.

2.3.   *Public sector research establishments*

The French "public research organisations" (PROs) are key actors in the research system. They conduct targeted research in support of public policies. The PROs increasingly interact with academic research at both the laboratory and individual scientist level.

The largest PROs are INSERM (medical research), INRA (agronomic research), CEA (nuclear research) and CNES (space research).

Together, the PROs represent a potential of 16 000 full-time researchers, about half the research potential of the higher education sector (excluding doctoral students) (Table 9). The PROs have a higher research potential than the engineering schools.

Table 9. **Scientific personnel at public research institutions, 1999**

	In full-time researcher equivalent[1] (in thousands)			
	Academic research human potential		PRO researchers[2]	Total human potential of public research sector
	Faculty (*enseignants chercheurs*) and CNRS researchers	Doctoral students (of which 3 000 postdocs)		
Exact and natural sciences	20.0	19.5	8.0	47.5
Engineering sciences	4.5	7.0	6.5	18.0
Social sciences and humanities	9.5	13.5	1.5	24.5
Total	34.0	40.0	16.0	90.0

1. Each faculty member counts as 0.5 persons in full-time research equivalent.
2. EPIC (*établissements publics à caractère industriel et commercial*) and EPA (*établissements publics à caractère administratif*) plus EPST (*établissements publics à caractère scientifique et technologique*), excluding CNRS (*Centre national de la recherche scientifique*).

2.4. Priority setting

National level of governance

The Interministerial Committee on S&T Research (CIRST – *Conseil Interministériel de la Recherche et de la Technologie*) is the decision-making and co-ordinating body at the government level. It meets once a year or once every two years to set the research priorities for the science system.

The Research and Technology Board (CSRT – *Conseil Supérieur de la Recherche et de la Technologie*) is an advisory body which annually assesses and makes recommendations on the R&D budget and research policy. Its members are drawn from the wider research system (including researchers' unions and industry).

The National Evaluation Committee for Universities (CNE – *Comité National d'Évaluation*) and the National Evaluation Committee for Research (CNER – *Comité National d'Évaluation de la Recherche*) are responsible for strategic evaluations of the public research institutions. They have a purely consultative role.

More detailed priority setting is established through four-year contracts (*contrats quadriennaux*) agreed between the Ministry of Research and each research institution, and five-year contracts agreed between the Ministry of Research, the regions and the research institutions (*contrats de plan État-régions*).

In addition, a number of inter-institutional co-ordination committees exist in fields of science which are important in several research organisations (life sciences, and social sciences and humanities).

Human and financial resources are decided on an annual basis by the Ministry of Finances and the Ministry of Research, and are negotiated with each institution. This relates not only to the general targeting of expenditure, but sets out the scientific activities and types of spending allowed within each institution.

Institutional level of governance (PROs and universities)

Each of the PROs and universities has its own governance structure, generally made up of a board and a scientific committee. They are thus able to define their own research priorities. Their options are nevertheless limited, since they have no control over the number of recruitment authorisations they receive each year, nor are they free to decide on the fields of science in which recruitment will take place.

Furthermore, about 80% of funding covers salaries, leaving little lee-way for changes of research orientation. This is all the more true since in the last few years, an increasing share of these institutions' "soft" funds has been managed directly by the Ministry of Research to fund various national programmes and establish direct linkages with the laboratories.

3.   Private sector R&D

In 1999, 4 650 firms reported R&D activities employing some 70 000 researchers. They performed R&D worth EUR 18.4 billion, or 60% of GERD. Private R&D performance has not substantially increased in value terms since the early 1990s, mainly owing to the decline in spending on military and R&D contracts related to "large technology programmes" (LTP).

Table 10. **Gross expenditure on business enterprise R&D, 1999**

EUR billion

Business enterprise RD	1995	1999
Current expenses	16.6	18.4
Expenses at 1995 prices	16.6	17.7
Evolution	100.0	107.0

SMEs (less than 500 employees) perform 21% of total industrial R&D, and this share seems to be on the rise. Half of the SMEs active in R&D are subsidiaries of multinationals. The electronics sector performs 24% of industrial R&D, followed by the aerospace, pharmaceutical and automobile sectors, with 10-14% of the total. The share of the aerospace sector is falling, while that of the pharmaceutical and automobile sectors is growing.

The major feature of industrial R&D in recent years has been the sharp decrease in public procurement contracts with an R&D component. This has affected not only the large firms which were the major producers of professional and military equipment, but also a significant number of small high-technology firms which were specialised suppliers for the large firms.

Table 11. **R&D expenditures by business enterprises, by sector, 1999**

Sector	%
Electronic and communication equipment	23.9
Pharmaceuticals	12.3
Automobile	12.3
Aerospace	13.7
Chemical industry	10.1
Energy	4.8
Machinery and equipment goods	7.8
Other	15.1
Total	100.0
Total (EUR billion)	18.4

4. Channels of industry-science relationships in France

Box 2. **Channels of industry-science relationships: a typology**

1. Contract research

1.1. Subcontracted research activities

1.2. Collaborative research co-financed by a firm

1.3. Collaborative research within a publicly financed programme

2. Consultancy and services

2.1. Transfer of expertise, know-how

2.2. Testing, technical services, access to specialised equipment

3. Intellectual property (IP) transactions

3.1. Build up of an IP portfolio

3.2. Licensing and transactions related to IP

3.3. Equity investment with or without IP implications

4. Knowledge spillovers and spin-offs

4.1. Provision of science park facilities (located near to campus)

4.2. Incubators (office space and services for start-up firms, located on campus)

4.3. firms' laboratories on campus

4.4. Public laboratories as lead users of innovative equipment

4.5. Informal interactions between public research staff and industrial researchers

5. Teaching/training

5.1. *Ad hoc* short courses for professionals/continuing education

5.2. *Ad hoc* diplomas for professionals/continuing education

5.3. Co-financing/sponsoring of postgraduate and doctoral students

6. Labour mobility

6.1. Public researchers staff taking up a position in industry

6.2. Industry research staff taking up a position in public research

6.3. Masters and PhDs with research techniques, methods and international networks going into industry

6.4. Joint laboratories between industry and public research

4.1. Research contracts, collaboration and consultancy

In volume terms, industrial funding of university research (at some EUR 100 million per year) represents about 2% of total expenditures on research, while the share of industrial contracts (at some EUR 400 million per year) in public sector research is about 8%.[2] This amount is equivalent to funding for European Framework Programme projects, and exceeds that of national and regional research contracts.

A number of policy instruments have been implemented to facilitate research collaboration between public and private institutions. Two of these deserve particular attention:

- *Joint laboratories between a firm and a public research institution.* The two parties contribute equally in terms of human, financial and infrastructural resources. Detailed provisions are made concerning intellectual property and publication of results. A few dozen joint laboratories have been set up in basic research, usually with large firms. In many cases, the contractual arrangement is simplified, amounting to a long-term contract between a firm and, say, a university, with the firm providing resources for contract work (but no permanent staff as such) and the university providing researchers' time, doctoral students and floor space in the university – these are known as "quasi-joint laboratories".

- *Technological research and innovation networks (réseaux de recherche et d'innovation technologique).* Established by the 1999 Law on innovation, these are large-scale, multi-year consortia of public and private research institutions oriented towards innovation and the promotion of start-ups in specific sectors/technological fields. These consortia provide partial funding for projects in the areas defined by their orientation committees; the committees select projects according to nationally defined criteria and procedures. There are currently 16 such large-scale consortia, with a total of EUR 200 million of funding in 2000.

- *National technological research centres (CNRT – Centres nationaux de recherche technologique).* Linked to the technological networks, these are regional competence centres which bring together existing industrial and public research capabilities on specific topics. The 13 CNRTs aim to facilitate and promote industry-science-education relationships and stimulate firm creation.

4.2. Commercialisation

Intellectual property rights (IPRs)

IPR practices differ among the public research organisations (PROs). CNRS leaves the industrial partner apply for the patent, and receives a percentage of the royalties. INRA and INSERM, on the other hand, generally prefer to retain ownership of the patent and to grant (often) exclusive rights to the industrial partner.

> **Box 3. The case of CNRS**
>
> In 2000, CNRS applied for 161 priority patents, compared with about 80 in the mid-1990s; in the same year 137 priority patents were extended to foreign patent systems.
>
> The patent portfolio of the CNRS consists of 982 priority patents, of which about 50% are co-owned. The total number of patents owned at least in part by CNRS was 4 181 in 2000.
>
> CNRS signs roughly 70 licences per year, with a total of 500 licences running (including software), generating about EUR 29 million in income per year. Of this, EUR 6.6 million are redistributed to the 130 laboratories concerned and another EUR 4.3 million to the inventors.
>
> One in three licences concern start-up firms, created with the aim of commercialising the patented technology.

Promotion of spin-offs, incubators, science parks

The 1999 Law on innovation pioneered in making the creation of technology-based firms a major objective of public research. The law aimed at:

- Providing incentives for researchers to participate in the creation of spin-off firms and facilitating the mobility of researchers to such firms (simplified leave-of-absence procedures for up to six years, right to hold shares and to be an executive in a spin-off firm, while retaining civil servant status, right to act as a paid consultant for the private sector, etc.).

- Simplifying the administrative procedures and providing fiscal incentives for would-be entrepreneurs, regardless of whether or not they belong to the public research sector.

- Enabling universities and public research organisations to set up incubators and introduce organisational innovations to better manage their relations with the private sector (*Service d'Activités Industrielles et Commerciales* – SAIC) and simplifying the creation of subsidiary firms specialised in early-stage financing and capital-risk.

- Providing funding for new technology-based firms, based on a nation-wide competitive bidding. A call for tender for spin-off firms has been launched every year since 1999 and has contributed to the financing of 500 new firms, at an annual cost of EUR 30.5 million. In parallel, ten seed-funding structures have been set up, combining public and private capital.

- Providing funding for the creation of incubators by public research institutions, based on a nation-wide competitive bidding. A call for tender was issued in 1999, and resulted in the creation of 31 incubators with a total public funding of EUR 45.7 million.

For a research institution to set up a subsidiary firm to manage capital-risk, seed funding and equity investments (capital-risk subsidiary) is a novel practice in France. Successful initiatives include:

- The technological university of Compiègne has set up Secant, jointly owned by banks and the private subsidiary of the university in charge of managing the contracts and the contract personnel.

- INRIA (*Institut National de la Recherche en Informatique et en Automatique*) has established INRIA-transfer.

5. Framework conditions for industry-science relationships in France

5.1. *Legal, institutional and regulatory context*

1982 Law on research (Loi d'orientation et de programmation de la recherche)

The basis for the French public research system was set up just after the second World War, with the establishment of CNRS, CEA and the launch of large technology programmes. The majority of the remaining research institutions and policy instruments (incentive funding, etc.) were created in the 1960s and early 1970s.

The cornerstone of the system is the 1982 Law on orientation of research *(Loi d'orientation et de programmation de la recherche* – LOP), as complemented in 1999 by Law on innovation (see above), which has the following main provisions:

- Overall co-ordination of civilian research funding (except for the universities) by the Ministry of Research through the civilian budget for R&D (*Budget Civil de Recherche Développement* – BCRD).

- Establishment of the Research and Technology Board (*Conseil Supérieur de la Recherche et de la Technologie* – CSRT), membership of which is made up by the "social partners" of research (representatives from various ministries, industry representatives, researchers' unions and organisations, academy of sciences, etc.).

- Mission statement to the effect that public research aims at knowledge production, education and transfers to society and industry.

- Definition of a role for the regions in research funding, provisions for public research institutions to set up private sector affiliates, creation of public joint venture research organisations (*Groupements d'Intérêt Publics* – GIP).

- Establishment of a framework for the evaluation of researchers at national level.

- Definition of a civil service status for all personnel (including researchers) in public research institutions (including universities), together with the prohibition of contract positions for scientists, except for doctoral students (this means, in practice, that it is almost impossible to hold a postdoctoral position in France).

5.2.   *Incentives and disincentives for industry-science relations*

Box 4. Incentive structure of industry-science relationships

1. The actors towards whom the incentives are oriented

1.1. Public sector researchers (salaries, promotion, conflict of interest regulations, mobility, etc.) and students

1.2. Managers of public research institutions authorities (financial autonomy, labour regulation, etc.)

1.3. Start-up firms (access to resources, tax treatment, etc.) and firms in general

1.4. Investors (*e.g.* tax policy)

2. The types of linkage towards which incentives are oriented

2.1. IP rights (who owns them? who pays? who benefits? what are the constraints?)

2.2. Research and consulting contracts (who benefits? what are the constraints?)

2.3. Public research infrastructure

2.4. Equity investments

3. The nature of the incentives

3.1. Regulatory

3.2. Financial

Evaluation of scientists and criteria for career advancement

Researchers and professors are evaluated and promoted by specialised scientific committees, of which two-thirds of the members are elected by the researchers of the given discipline. These committees are autonomous and their recommendations are generally implemented by the administrative authorities.

This gives strong weight to the existing disciplinary structure and leads to an "egalitarian" style of evaluation. It has been argued that the evaluation committees are not favourable to researchers having strong industry ties:

- At best, such ties are ignored, which may make it difficult for the researcher to compete with colleagues dedicated to purely academic activities.

- At worst, they are considered to generate income for the researcher, making his/her promotion unnecessary.

This situation is slowly changing as the new innovation culture spreads through the research institutions and scientists are encouraged to enhance their interactions with business.

Incentives for firm creation by researchers

The 1999 Law on innovation made a significant breakthrough in this area but it is too early to evaluate its real impact.

Researchers' income from industrial property

Researchers who are civil servants can benefit from their contribution to patenting in the following way: funds paid to individual inventors in a patent amount to 50% of the total royalties paid to the institution. However, this rule has certain limitations:

- The researcher is the sole beneficiary; the benefits cannot be passed on to his/her successors.

- IP rights are granted only to the direct inventors.

- The inventors must be civil servants (not doctoral students or postdocs).

It has been proposed that the income from intellectual property and contracts should be placed in a "fund" for subsequent redistribution to all those who contributed to the innovation in one way or another.

5.3. *Organisational arrangements*

Box 5. Organisational arrangements for managing ISRs

1. Direct management of the linkages by the researchers themselves.

2. Service of the research institution managing the linkages (*e.g.* technology transfer office).

3. Subsidiary of the research institution managing the linkages (technology transfer subsidiary).

4. Subsidiary firm of the research institution managing capital-risk/seed funding and equity investment (capital-risk subsidiary).

5. Linkages managed by external independent organisations, either public or private.

Direct management of the linkages by the researchers themselves

In the social sciences (management, law, etc.), in the majority of the cases, the researchers themselves manage the linkages.

This is also frequently the case in the other disciplines, since civil servants can spend 20% of their time as consultants, without any limit in terms of income, if the contractor is a private institution. In practice, since researchers seldom request formal authorisation for consultancy contracts, there are no meaningful statistics on such activities.

For the management of individual contracts, researchers usually have to establish themselves legally as consultants, a step that involves significant administrative and fiscal constraints. Therefore, in many cases, consultants arrange for their contracts to be managed by an "association"[3] which in turn pays them a salary. However, this may be illegal if the contract to be managed is with a public body.

It is difficult to draw the line between "personal" consulting activities (which must not use laboratory resources and equipment) and contracts carried out through and to the benefit of the laboratory. This raises a number of complex ethical questions. The borderline between public and private activities is all the more blurred as medical researchers are officially allowed to hold private practices (within certain limits) in public hospitals.

Service of the research institution managing the linkages

All public research institutions and most universities have created a service to take charge of industrial relations. The role of such industrial relations services vary: in many cases, they are responsible for managing contracts and employing the required dedicated personnel. Such tasks are not easy to handle, because the service has to address the following problems:

- *Recruitment of the dedicated personnel.* It is now possible for the public research organisations (PROs) to hire personnel on a fixed-term contractual basis. However, there are a number of restrictions which make the procedure rather heavy; hence, as noted above, there is a tendency to create "associations" to provide the necessary flexibility for handling the contract.[4]

- *Financial incentives for the researchers responsible for the invention.* Here again, there are different interpretations of the rules and different practices in terms of both the financial flows regarding the work itself and those regarding possible intellectual property aspects.

The Ministry of Finance favours organisational arrangements where all the formalities are carried out by a service within the university or PRO since this facilitates financial control. The other side of the coin is the heavy and bureaucratic character of the management. Furthermore, any disputes arising with contracting firms must be settled by administrative jurisdictions, a procedure which firms dislike. In general, neither clients nor researchers are in favour of this scheme.

In order to facilitate the direct handling of industrial relations by the universities themselves, the 1999 Law on innovation enables the creation of "industrial and commercial activity services" (*Services d'Activités Industrielles et Commerciales* – SAIC), which benefit from somewhat simpler procedures for hiring personnel and managing contracts. However, the implementation of the SAICs is not yet possible since complex issues related to taxation and to public management of private funds and non-civil service personnel have yet to be sorted out.[5]

In practice, the model according to which industry-science relationships are handled by the public research organisation itself does not work. In general, the central "commercialisation service" co-exists along with a number of "associations" which are not formally linked to (or sometimes even known by) the university or PRO. These associations manage contracts and personnel for the laboratories and/or individuals.

Management of the linkages by a subsidiary of the research institution

Another model is a dual scheme with:

- A service within the university or PRO which takes on the co-ordination role (but not the management-operational role).

- A private (profit or non-profit) official subsidiary of the university or PRO with an operational role (managing the contracts and hiring contractual personnel).

i) The subsidiary is a private institution, be it a firm (SA) or an association (non-profit institution):

A "framework contract" is drawn up between the subsidiary and the university: the subsidiary is autonomous and signs the contracts; the university executes (most of) the contracts. University researchers are not paid, but their laboratory receives income from the subsidiary.

This scheme is accepted reluctantly by the authorities in cases where the subsidiary is a private association or firm. So far, however, the authorities have authorised contracts with private subsidiaries provided that the relationships between the university/PRO and its subsidiary are clearly defined.

Table 15. **Examples of private subsidiaries**

Higher education institution	Subsidiary firm for commercialisation of technology activities
Technical university of Compiègne	Gradient
Lyon 1 university	Ezus
École normale supérieure de Cachan	Science Pratique
École des Mines	Armines
INSA Lyon	Insavalor
École centrale de Lyon	Lyon innovation
CNRS	FIST

ii) The subsidiary is a co-operative structure:

- *Groupements d'Intérêt Public (GIP) (public interest groups).* This new category of institution was created by the law of 15 July 1982 to enable public research organisations to "conduct, together and with private institutions, during a given period, research and development activities, or manage installations of common interest needed for these activities". The GIP were created to foster the emergence of "networks of actors". The dozens of GIP that have been created since 1982 have had various objectives: exploiting INRA's expertise on seeds, joint development of an innovation, assistance to a group of firms with a common project, etc. Some of the GIP have since been transformed into private firms, others have ceased their activities; 36 GIP were active in 1998. The main problem with GIPs is that they are not allowed to to hire specific personnel (staff have to be hired by the member institutions and then devolved to the GIP).

- *Groupements d'Intérêt Economique (GIE) (economic interest groups).* Created in 1967, GIE are private institutions in which PROs can participate. A GIE has commercial objectives, it is simple to set up and can hire personnel. However, the authorities are often reluctant to authorise their establishment. PROs need formal governmental approval for all institutional participation in an outside structure, be it public or private. The authorities prefer the PRO to handle contracts internally and have made it difficult to develop the subsidiary model. Therefore, it is not surprising that there are less than 30 industrial subsidiaries of PROs. One of the effects of the Law on innovation may be that the authorities will no longer allow the creation of private subsidiaries, preferring the SAIC model described above. It remains to be seen whether the SAIC are, in fact, as simple to create and manage as has been claimed. This situation could clearly give rise to a situation in which ISR are made more complicated and more bureaucratic than before.

Box 6. "Industrial" subsidiaries of PROs and universities

Some PROs have industrial subsidiaries, although the general trend is for the PROs to decrease their financial and other involvement in projects of this kind. The reasons are that these subsidiaries tend to engage in activities that are too far removed from the "cultural base" of the parent organisation and that they have so far yielded only modest economic benefits. The major PROs often have less than ten industrial subsidiaries, many of which are small.

Some of these subsidiaries do not behave like private firms and have public service objectives. For example, INRA (the national agronomic research organisation) has set up "Agri-obtentions" to commercialise the seeds developed by INRA. This company has a public service mission since it sells a very large range of seeds, many of which serve markets which are too small to be profitable.

Two PROs are in a different situation: the Commissariat à l'Énergie Atomique (CEA), with its large network of industrial participants, including significant shares in large companies (Framatome and Cogema); and the CNES, with its important participation in Arianespace.

Linkages managed by external independent organisations

This type of organisational arrangement is found at the regional level. A number of institutions provide technological expertise and services to SMEs. These institutions usually rely for their scientific expertise on the personnel of an university, an engineering school, a PRO laboratory, or on a Centre for industrial techniques (*Centre Technique Industriel* – CTI).

In order facilitate the use by SMEs of such technology transfer infrastructure and to identify the best targets for public financial support, a "label" process has been set up: a national commission evaluates institutions seeking to obtain the status of Technological Resource Centres (*Centres de Ressources Technologiques* – CRT). The "label" is awarded based on criteria relating to scientific and technological capabilities as well as professional qualifications. Today, 21 such centres have been granted the CRT status.

Furthermore, a Technology Diffusion Network (*Réseau des Conseillers en Développement Technologique* – CDT) has been set up in each region. It is made up of technology management experts from: Chambers of Commerce; PROs such as the National Agency for Innovation (*Agence Nationale pour l'Innovation* – ANVAR), and the CEA; regional representatives of the Ministries of Research and Industry, etc. They do not carry out contract work, but rather conduct initial diagnostics of SMEs and recommend qualified service providers; they can also provide advice to firms on how to benefit from public co-financing schemes.

6.   Human resources and ISRs in France

6.1.   Employment of graduates

In science, technology and engineering, universities are not the most important component of the French higher education system. The so-called *Grandes Écoles* have a quasi-monopoly for the delivery of the title of "engineer", which is granted each year to about 22 000 students – twice as high as the total number of PhDs. The curriculum of the *Grandes Écoles* is completely independent from the university system. Registration is based on a specific entrance examination, prepared in two years in "preparatory classes", which are not university-based. After obtention of their diploma, many engineers go on to obtain a PhD in engineering sciences, physics or chemistry.

In agronomy, forestry and food sciences, the *Écoles d'agronomie* (similar to the *Grandes Écoles* in their principle and organisation) play a key role. The major business schools (*Écoles de Commerce*) are also outside the university system. Even architecture, music and the arts have their own *Écoles*, which are not universities. All of these *Écoles* deliver diplomas but not PhDs, for which the universities have a monopoly.

Table 16. Employment of PhDs 18 months after obtaining their diploma, 1998

Percentages

Permanent position in research or higher education	20:3
Contract position in a public research institution, in France or abroad	20.7
Employment in a firm	17.0
Employment in an administration	4.4
Teaching in a secondary school	5.2
Unemployed	6.7
Other – return to country of origin, situation unknown	25.7
Total	100.0
Number of PhDs delivered	11 000

Note: About 25% of PhDs delivered in France are delivered to non-French citizens; about one-third of all PhDs are produced in the social sciences and the humanities.

Some of these *Écoles* have a monopoly over the diplomas in their field (there is no university curriculum nor PhD in the field); others co-exist with university curricula, but diplomas issued by the *Écoles* are usually viewed as having a higher professional value.

Most of the students interested in engaging in doctoral studies generally enter university at the PhD level, having obtained their diploma from the *École*. In France, the terms "university" and "PhD" are used differently than in many other countries. PhDs in France are mainly trained for future employment in public institutions and universities; as shown in Table 15, 18 months after their diploma, only 17% of PhDs work in industry. Even accounting for postdocs and researchers for whom no information is available, only about one-quarter of all French PhDs will end up working in a firm.

Regarding the linkages between higher education institutes and public research, the *Écoles* traditionally maintain close contacts with the professional world, while the universities have closer contacts with academic research. There are thus two possibilities for improving the linkages, both of which are actively pursued:

- The universities should tighten their relations with the industry circles.

- The *Écoles* should develop closer links with the university laboratories and/or relevant public research institutions, and enhance their research activities through these linkages.

6.2. *Policy mechanisms to reinforce mutually training and industry-science relationships*

Internships for graduate students in SMEs

Four- to six-month internships in a SME for students working closely with a professor from a "competence centre" of the higher education institution. Some financing is possible.

Postdoctoral positions in SMEs

Financing of 50% of the cost of employment in a SME for a maximum of two years for a young PhD without industrial experience.

Employment in research in SMEs

Financing of 50% of the cost of employment in research in a SME for a maximum of two years for a young graduate or engineer.

Innovation project prepared by a higher education institution and a SME

Financing of 50% of the cost of a technological innovation project prepared by a group of students with a SME.

Research grant for preparing a PhD in a firm (Contrat Industriel de Formation par la Recherche en Entreprise – CIFRE)

Financing of 50% of the cost of employment of a doctoral student (for a period of three years). The student performs the research under the supervision of a university professor, on a topic chosen jointly by the firm and the professor.

Research grant for a research project for a technician in a SME

Financial support to a technician performing a project for a SME, with the support of a higher education institution.

6.3. *Mobility of research personnel*

This mobility is extremely low. Each year, some 40 scientists (from a total of 25 000) leave their public research organisation to work in a firm (0.16%). This figure rises to about 150 when temporary leave is taken into account (0.6%). A similar number of industrial researchers (140) are assigned to PROs on a temporary basis.

7. Concluding remarks

Industry science relationships in France are shaped by some key features of the higher education-public research landscape:

- Public research is carried out in a variety of institutions: universities, *Grandes Écoles* and public research organisations (PROs). There are two major types of PRO: EPST (generally more fundamental research, with no obligation to employ contractual resources); and EPIC (generally more applied research, with an obligation to employ 20-40% contractual resources).

- In higher education, the *Grandes Écoles* play a significant role in research, particularly in engineering sciences.

- CNRS personnel employed in full-time research are almost totally located on university campuses, usually in "mixed" laboratories, *i.e.* laboratories staffed by both university and CNRS personnel.

- The vast majority of public researchers, be it university or PRO personnel, has civil servant status. This implies nationally defined, harmonised and rather rigid rules regarding salaries, career development and other aspect of the incentive structure.

- PRO laboratories ("targeted research public laboratories") are increasingly merging their resources with university laboratories, thus generalising the notion of "mixed" laboratories.

- All universities are public and have the same status. Staffing and resources are negotiated and decided at the national level (part of which is agreed in the framework of a four-year contract). Recruitment and promotion procedures have mixed local-national components.

In that context, it would be unrealistic to seek the improvement of ISRs by simply emulating successful experiences of countries with rather different institutional settings. The efficiency of ISRs reflects that of the public-private interface. In the French context, this interface is not clearly and efficiently dealt with and here some lessons from foreign practices could be learned. Recruitment of contract researchers, financial rewards for faculty members, the handling of IPRs all involve complex, lengthy and sometime inconsistent procedures, with major discrepancies between what the government, especially the Ministry of Finance, wants to achieve and what goes on in practice.

The 1999 Law on innovation is certainly a step in the right direction, but the legal and regulatory stalemates which prevent some of its provisions being implemented show that it falls short of removing all impediments to fruitful ISRs.

Industry-science relationships have been the subject of a successive layers of initiatives over the last ten to 15 years. Taken individually, each initiative has its merits, but the combined and systemic impact of all initiatives has never been evaluated.

One prerequisite is to make ISRs in France more transparent to policy makers. The visibility of existing initiatives is all the more limited since their implementation methods, their names, the conditions of application vary from one university to another, from one research institution to another. This fragmentation presents a major policy concern.

Priority should be given to the monitoring and evaluation of the impacts of the 1999 Law on innovation, including the influence of other framework conditions on outcomes.

NOTES

1. Public research organisations (PROs), whose statutes specify a particular objective or sector for their research activities.

2. The CEA alone accounts for 40% of industrial contracts for public research; this figures includes contractual relationships with its industrial subsidiaries.

3. These are non-profit structures which are private and extremely easy and cheap to set up. Of course, they have to abide by the labour regulations and, since they are created with no capital, the persons in charge, in particular the Chairman of the Board and the Treasurer, are personally responsible for the financial situation.

4. Hence, also, the problem of reliability of the statistics on contracts provided by the administration of the PROs; it is well known that a very significant part of the contracts do not appear in the statistics, since they are not made with the laboratories themselves but with parallel "associations", whose accounting records are not accessible by the PRO administration.

5. In practice, the decrees (*décrets d'application*) have yet to be published by the government.

BIBLIOGRAPHY

Amable, Bruno, Rémi Barré and Robert Boyer (1997), *Les systèmes d'innovation à l'ère de la globalisation*, Economica, Paris.

Barré, Rémi (ed.) (1999), *Science et technologie – indicateurs*, Economica, Paris.

Callon, Michel (1998), "Des différentes formes de démocratie technique", *Annales des Mines*, January, Paris.

French Ministry of Finance (1999), *Projet de Loi de Finances pour 2000 : État de la recherche et du développement technologique*, Imprimerie nationale, Paris.

Latour, Bruno (1999), *Politiques de la nature – pour faire entrer les sciences en démocratie*, La Découverte, Paris.

Majoie, Bernard (Rapport du groupe présidé par) (1999), *Recherche et innovation : la France dans la compétition mondiale,* Commissariat Général du plan, La Documentation française, Paris.

OECD (1998), *STI Review, No 22, Special Issue on "New Rationale and Approaches in Technology and Innovation Policy"*, OECD, Paris.

Chapter 5

INDUSTRY-SCIENCE RELATIONSHIPS IN THE UNITED KINGDOM[*]

1. General introduction

The relationship between science and economic success has been a preoccupation for the United Kingdom's science and innovation policy for more than a century (Georghiou, 2001). In the past decade, the desire to improve industry-science linkages has become the central organising principle for science-funding initiatives, encompassing issues such as prioritisation and network-building through foresight; a "customer-focus" for research funding bodies; and several initiatives aiming to foster entrepreneurship among faculty and graduates. This chapter aims to present the current situation of industry-science relations in a format which enables international comparison and benchmarking of policies. Many of the initiatives described here were inspired or at least informed by the experiences of other countries. One could cite the experience of Japan in technology foresight and of the United States in commercialisation of technologies as two such examples. On the other hand, the United Kingdom is itself the subject of numerous international visits by delegations seeking to learn from its experiences in areas such as privatisation and the operation of market principles in research. This interest would not exist were it not for the sustained successes of the UK science base in terms both of quality and of establishing linkages with industry. Despite the substantial learning and policy-transfer which has taken place, until recently there has not been a systematic means to make comparisons. In this report, the UK system and experience are described in a way which is intended to facilitate this process.

By way of introduction, the knowledge-production capacity of the United Kingdom is described, together with the policy system which supports it. Industry-science relations are described in three main spheres, those of research, collaboration and commercialisation. In line with the recent European Union benchmarking project in this area, there is then a discussion of the influence of framework conditions, encompassing the legislative background and the role of incentives and disincentives. Prior to the concluding discussion, a separate section is devoted to the issue of human-capital production and mobility, often cited by industry as the most important contribution that science can make.

[*] This chapter has been written by John Rigby and Luke Georghiou, PREST, University of Manchester, United Kingdom. The authors wish to thank the following experts or organisations that they contacted during the writing of this study: Association of Independent Research and Technology Organisations; AURIL (Association of University Liaison Officers); Biotechnology and Biological Sciences Research Council; Confederation of British Industry; Chemical Industries Association; Dr Jeremy Howells; Department of Trade and Industry; London School of Economics; Manchester Innovation Ltd; Medical Research Council; Science Policy Research Unit.

1.1. Methodological note

The report synthesises a number of forms of data – from interviews, specialist desk research and statistical measures from national and international reports. Statistical research informs the first two main sections of the report. Interviews with major policy makers in the UK research system and desk research mainly involving official reports and Web sites provide the basis of Sections 5 and 6.

The report also draws heavily upon two recent studies by PREST:

- The most recent national survey in the United Kingdom: Howells, J., M. Nedeva and L. Georghiou (1998), "Industry-Academic Links in the UK", Report to the Higher Education Funding Councils for England, Scotland and Wales.

- Cox, D., L. Georghiou and A. Salazar (2000), "Links to the Science Base of the Information Technology and Biotechnology Industries", Report produced on behalf of the Economic and Social Research Council for the Director General of Research Councils, http://les1.man.ac.uk/PREST.

The report is based on work carried out by the authors for a study under the auspices of the European Commission project to benchmark framework conditions in industry-science relations across the Union.

2. Knowledge production in the United Kingdom

2.1. Institutional and funding structures

This section examines the sites and forms of knowledge production in the context of the United Kingdom. In particular, it examines the following issues: the institutions and structures of public research; the prevailing research culture within public research; the structure of private R&D; and the funding sources available to public, private and other institutions. This section therefore concentrates on the actors in the system and the central institutions responsible for research. The overall context is that the United Kingdom's gross expenditure on R&D (GERD) was GBP 15.5 billion in 1998, equivalent to 1.8% of GDP. In real terms this has fallen from 2.2% in 1985 and places the United Kingdom below the OECD average and just above the EU average. However, knowledge-based industries provide 51% of business sector value added and grew at 4.1% per annum in the period 1985-96 (both figures place the United Kingdom fourth highest among OECD countries, according to the *OECD Science and Technology and Industry Outlook 2000*). Scientific performance is also strong. With 1% of the world's population, the United Kingdom funds 4.5% of the world's science, produces 8% of the world's scientific papers and receives 9% of citations.

Figure 1 shows the main actors involved in the funding and performance of research in the United Kingdom, while Figure 2 shows the flow of funding between them. Key features to note are that industry is the largest performer of R&D, accounting for 66% of the total, followed by universities with 20% and government laboratories with 13%. Industry is also the largest funder of R&D, spent mainly in its own premises but also accounting for 7.27% of higher education research income. Defence remains a large sector of the economy. While civil R&D was GBP 13 207 million in 1998, defence R&D was GBP 2 346 million in the same period.

Figure 1. **UK funding and performance of R&D**

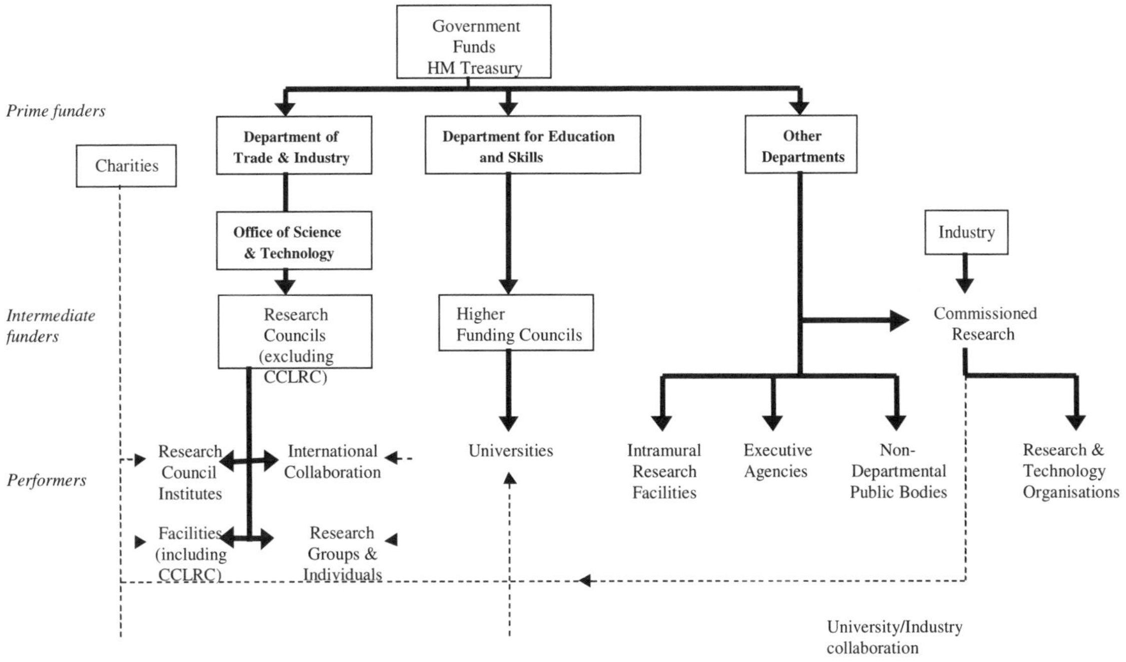

Note: The Department for Education and Employment has been reorganised following the general election of 2001 and is now known as the Department for Education and Skills.
Source: Cunningham and Hinder (1999).

Universities are funded for research through what is known as the dual support system. Higher Education Funding Councils (separate bodies for England, Scotland, Wales and Northern Ireland, with funds derived from ministries responsible for education) provide general funding, used mainly for academic salaries and research infrastructure, while Research Councils [with funds derived from the Office of Science and Technology (OST) in the Department of Trade and Industry (DTI)] provide funding for projects (including salaries of contract researchers), research training and centres on a competitive peer-reviewed basis. Some of the Research Councils also operate their own institutes. The other principal funding source for research is the charitable, non-profit sector, notably the Wellcome Trust which is the largest single funder of medical research. Universities and Research Council institutes collectively form what is generally referred to as the "science base".

2.2. *Public research: the higher education sector*

Universities are the main performers of basic research in the United Kingdom. Although principally funded by government, they are independent institutions with charitable status. Their employees are not civil servants. At August 2000 there were 114 university institutions in the United Kingdom, counting separately the colleges of the federal universities of London and Wales. Older universities in a few cases are centuries old, with civic universities founded in the late 19th and early 20th centuries and most of the rest in the 1950s and 1960s. Former polytechnics were given the status of universities in 1992 and are often referred to as "new" universities.

There were 1 846 000 students in higher education in 1998/99, of whom 22% were postgraduate and 12% from outside the United Kingdom. Within the higher education sector, research is carried out in the vast majority of institutions, although there is a strong concentration of research activities in

111

around 45 of the largest and older universities. Furthermore, the five largest institutions by research income receive 25% of all research income, and the top 15 receive about half of this income. The bottom 50% of institutions account for under 10% of research funding. This pattern applies for all research indicators and funding sources.

Figure 2. **The flow of funds for UK R&D, 1998–99**

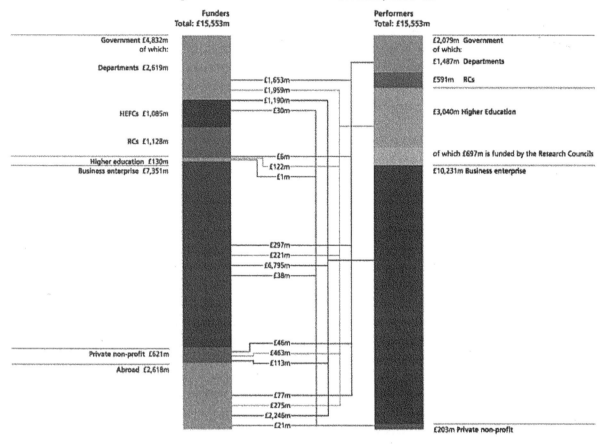

Source: Department of Trade and Industry (2000b).

The Funding Councils' support of research is mostly aimed at that which is curiosity based. Indeed, all Funding Council research is nominally considered to match the basic category of the *Frascati* definition. The quality of this research output is assessed through the Research Assessment Exercise (RAE), which takes the form of a disciplinary panel-based review of publications and other data and which occurs every four or five years. The results of this exercise (on a seven-point rating scale) are used to allocate the vast majority of the funds provided by the Funding Councils. This quality-related or QR funding is allocated to reflect the quality and volume of research at institutions in different subjects and amounts to around GBP 743 million a year. Funds for the supervision of research students are GBP 65.6 million.

Generic research (GR) funding has been provided by the Higher Education Funding Council for England in response to the theme of wealth creation in the 1993 Science and Technology White Paper, *Realising Our Potential* (OST, 1993). GR funding recognises collaborative research that does not have a single beneficiary and supports research in the university system where universities retain the intellectual property of the research and the publication rights. In 1999-2000, GR funding amounted to GBP 20 million. GR funding is aimed to support university links with industry. However, this funding stream is now being terminated with the introduction of new incentive funding for industrial collaboration (see Section 5).

As noted above, Funding Councils for each of the countries that comprise the United Kingdom have existed for some years. As the devolution of political power to the new national assemblies of Scotland, Northern Ireland and Wales gathers pace, national policies for research have already begun to vary as Funding Councils react to local needs and priorities. In Scotland, for example, while a significant proportion of Funding Council support for research still comes through the Research Assessment Exercise (85% in 1999), the new Research Development Grant scheme (RDG) allocates around 10% of Funding Council grants with the aim of bringing about a realignment of the research base with perceived economic and social priorities (Table 1).

Table 1. **Scottish Higher Education Funding Council Research Funding, 1999-2000**

Funding stream (1999-2000)	Funding (GBP thousand)	%	Means of allocation
RAE-based grant	110 134	85.3%	Competitive formula
Research development grant	10 000	7.7%	Submission-based, with emphasis on collaboration.
Initiative funding	3 085	2.4%	Various
Contract research staff			Various
Promotion of commercialisation			Formulaically against a quality threshold
Professionalisation of commercialisation			Various
UK Activity Funding	5 950	4.6%	Competitive bidding
Joint Research Equipment Initiative (JREI)			Competitive bidding
Humanities Fellowships			Various
Arts and Humanities Research Board (AHRB)			Competitive bidding
UK Research Strategies Libraries Programme			Competitive bidding
Total funding for research	129 159	100%	

Source: Scottish Higher Education Funding Council (2001).

The six Research Councils are non-departmental public bodies with budgets from the office of the Director-General Research Councils in the Office of Science and Technology. They are the:

- Biotechnology and Biological Sciences Research Council (BBSRC).

- Economic and Social Research Council (ESRC).

- Engineering and Physical Sciences Research Council (EPSRC).

- Medical Research Council (MRC).

- Natural Environment Research Council (NERC).

- Particle Physics and Astronomy Research Council (PPARC).

In addition, the Council for the Central Laboratory of the Research Councils operates large central facilities. Three councils – MRC, NERC and BBSRC – operate institutes or research units of their own as well as supporting university research through grant funding. PPARC also handles UK contributions to international facilities in particle physics and astronomy.

Table 2 shows the classification of government research carried out by the Funding and Research Councils and the budget in 1998-99 for each Research Council.

Table 2. **UK Government science budget by *Frascati* definition, 1998-99**

Science budget	Basic pure %	Basic orientated %	Strategic %	Applied specific %	Experimental development %	Total %	Total GBP million
OST – DTI	0	80	12	8	0	100	28.3
BBSRC	0	53	39	8	0	100	184.1
ESRC	15	27	41	17	0	100	63.6
MRC	0	62	28	10	0	100	275.8
NERC	3	47	33	11	6	100	160
EPSRC	18	32	46	4	0	100	350.6
PPARC	94	0	5	0.3	0	100	204.9
Total science budget	21	39	32	7	1	100	1 267.4
Higher Education Funding Councils	100	0	0	0	0	100	1 085.0

Source: Department of Trade and Industry (2000b).

Table 3 shows the research intensity of each subject field (using cost centres) by indicating the amount of research income in 1998 and the change in this sum in the period 1995-98. This emphasises the United Kingdom's strong and growing investment in biomedical research. A notable feature is the apparent decline in funding for information technology research, although this is mainly the result of re-coding of research expenditure by the Higher Education Statistics Agency (HESA).

HESA maintains databases of information about the research activities of academic and research staff. These databases have been examined to establish the proportion of staff working in different subject areas (termed "fields of science" in the ISCED-97 definition), and the rates of growth and decline in staff numbers working in them. The fields of science have been aggregated to create five main subject areas:

- Humanities, religion and theology, fine and applied arts.

- Social and behavioural sciences, law, commercial and business administration, mass communication and documentation.

- Natural science, mathematics and computer science.

- Medical science and health-related.

- Engineering, architecture and town planning, transport and communications, agricultural sciences.

Table 3. **UK HE sector research income by cost centre in 1998 and growth, 1995-98**

Cost centre	Funding (GBP thousand)	Growth (%)
1 Clinical medicine	523 758	16.4
10 Biosciences	265 070	19.3
12 Physics	115 556	4.4
11 Chemistry	93 848	10.3
20 Electrical, electronic and computer engineering	79 088	24.2
29 Social studies	73 432	7.1
21 Mechanical, aero and production engineering	72 961	1.2
14 Earth, marine and environmental sciences	48 984	13.5
16 General engineering	46 790	21.4
4 Anatomy and physiology	40 111	-0.1
13 Agriculture and forestry	39 172	31.0
18 Mineral, metallurgy and materials engineering	36 858	3.3
27 Business and management studies	36 398	21.6
7 Psychology and behavioural sciences	31 520	8.3
3 Veterinary science	25 574	7.5
19 Civil engineering	25 387	-1.6
24 Mathematics	22 933	3.0
17 Chemical engineering	20 866	9.2
9 Pharmacology	20 522	27.4
8 Pharmacy	20 368	24.7
23 Architecture, built environment and planning	18 531	28.7
28 Geography	18 427	11.5
6 Health and community studies	15 909	30.3
25 Information technology and systems sciences	14 336	-25.8
32 Humanities	13 800	20.1
5 Nursing and paramedical studies	12 048	48.7
2 Clinical dentistry	11 718	13.5
33 Design and creative arts	5 930	51.6
31 Language-based studies	4 921	16.8
30 Librarianship, communication and media studies	2 025	3.6
15 General sciences	961	-2.0
26 Catering and hospitality management	657	10.6
22 Other technologies	153	-62.4

Source: PREST, based on HESA data (2000).

The following tables show the different numbers of research scientists working in these five different fields and how the numbers of scientists engaged in each field have varied over the last five years. HESA groups academic staff into three categories: teaching only, research only, and research and teaching. Table 4 shows the total number of scientists engaged in both research only and teaching and research, *i.e.* categories two and three.

Table 4. **Total research scientists in the higher education sector**

Field of science	1994-95	1995-96	1996-97	1997-98	1998-99
Humanities	11 507	12 022	12 151	12 219	12 444
Social sciences	21 552	22 578	22 710	22 277	23 372
Natural sciences	21 065	21 888	21 840	20 705	20 327
Medical sciences	17 598	20 253	21 570	21 793	22 555
Engineering sciences	15 780	15 846	16 152	15 028	15 204
All others, etc.	2 270	1 723	1 709	3 349	4 452
Total	**89 773**	**94 310**	**96 132**	**95 371**	**98 354**

Source: HESA, ad hoc enquiry (2000a).

The number of researchers increased in the period 1994-98 by 9.6% including the "All others" group. When the "All others" category is excluded, the rate of increase of researchers staff falls to 7.3%. The rates of increase are uneven, with some actual declines. In natural sciences and engineering there are fewer researchers at the end of the period than at the beginning. In the social sciences, the number of researchers rises by 8.4% and in medical sciences by 28%.

The number of research only staff is shown in Table 5 below. The numbers of university staff who are solely engaged in research has increased over the period by around 9.8%, a similar increase to the growth in the size of the overall research total. The high proportion of these staff (29%) is caused by a large expansion in recent years in the numbers of contract researchers in universities. Variations in the rate of growth of staff numbers by subject are similar to those found in the rate of growth of the total number of research active staff. When the number of dedicated research staff is counted in proportion to the total number of research staff (which includes staff engaged in both teaching and research), the proportion for all areas of science does not change over the period.

Table 5. **Total university research only staff**

Field of science	1994-95	1995-96	1996-97	1997-98	1998-99
Humanities	716	834	852	862	930
Social sciences	3 017	3 428	3 401	3 195	3 225
Natural sciences	8 522	8 954	8 992	8 835	8 732
Medical sciences	7 379	8 197	8 708	8 635	9 005
Engineering sciences	5 613	5 702	5 734	5 509	5 500
All others, etc.	390	251	245	643	765
Total	**25 637**	**27 364**	**27 933**	**27 678**	**28 158**

Source: HESA, data for PREST (2000).

2.3. *Public sector research establishments*

The government laboratories sector – which is now termed the public sector research establishments sector (PSREs) – has been considerably reduced in recent years through the privatisation of government laboratories (Boden *et al.*, 2001). Civil spending on R&D by government departments has declined, but remains substantial. It is now disbursed primarily on a competitive basis, with former government laboratories remaining major performers.

Remaining PSREs exist to assist in the pursuit of government objectives, including:

- Improving quality of life (*e.g.* medical research).

- Economic development through advance in basic science (*e.g.* agricultural research).

- Informing government policy making (*e.g.* advice on genetically modified crops).

- Statutory scientific testing and regulatory functions (*e.g.* testing animals for disease) (Baker, 1999).

These organisations fall into two main groups:

- Departmental bodies, responsible to central government departments, either as an executive agency or as part of the department.

- Research Council institutes, accountable to the Council concerned.

The majority have between 100 and 1 000 employees, with some Medical Research Council establishments having only 20 employees. However, at the other extreme, the Defence, Evaluation Research Agency (DERA) had over 10 000 employees and a budget of GBP 1 billion per year prior to a partial privatisation (public-private partnership) in 2001 which created the Defence Science Technology Laboratory with around 3 000 employees located within the Ministry of Defence and a new public-private partnership defence research contractor, QinetiQ plc, wholly owned by government but intended to undergo further privatisation with the remaining employees.

A large number of scientists work in the PSREs. The vast majority of these research staff are engineers and scientists, although a very small number of researchers are qualified in either social science or the humanities (4.4%). Data on the specific scientific field of these scientists are not available although the following, given in Table 6, is known. Reduction in numbers due to budget cuts and privatisation is evident. For example, the number of researchers employed by the Department of Environment, Transport and the Regions[1] has fallen to 15% of the number in the predecessor ministries (DoE and DoT) because of the privatisation of the Building Research Establishment and the Transport and Road Research Laboratory. However, the scientists who continue to work in these establishments are still primarily engaged in providing services to government, in some cases in pursuit of an industry-science relationship (ISR) mission.

2.4. *Priority setting*

In the UK system, each Research Council establishes its own research priorities within the context of overall priorities set by the Director General of the Research Councils and informed by the Foresight Programme. The Director General of the Research Councils is himself part of the process of priority setting through links with the government's Chief Scientific Adviser, the Council for Science and Technology, which is chaired by the Secretary of State for Trade and Industry, and the Chief Scientific Advisers in each government department. An informal ministerial committee, the Ministerial Science Group, also contributes to the setting of priorities, although its main purpose is to co-ordinate science policy across government.

Despite the presence of priorities and strategic programmes, the majority of research carried out in universities is funded in "responsive mode", that is to say the topic and approach are identified and proposed by the researcher and there is no restriction on publication other than the normal governance of peer review.

Table 6. **Total personnel engaged on R&D within government, by department, 1989–90 to 1998–99**

Organisation	Acronym	1989–90	1998–99
Research Councils			
AFRC		4 211	–
BBSRC		–	3 241
ESRC		111	95
MRC		3 263	2 873
NERC		2 720	2 627
SERC		2 729	–
EPSRC		–	291
PPARC		–	299
CCLRC		–	1 691
TOTAL RESEARCH COUNCILS		13 034	11 117
Civil Departments			
Ministry of Agriculture, Fisheries and Food	MAFF	1 801	1 875
Department for Education	DFE	12	–
Employment Department	ED	62	–
Department for Education and Employment	DfEE	–	189
Manpower Services Commission/ Training Agency	MSC/TA	137	–
Department of the Environment	DOE	890	–
Department of Transport	DOT	614	–
Department of Environment, Transport and the Regions	DETR	–	223
Department of Health and Social Security	DHSS	–	–
Department of Health	DH	628	698
Health and Safety Commission	HSC	98	123
Home Office	HO	251	388
Department of Culture	DCMS(formerly DNH)	–	223
Department for International Development	DFID (formerly ODA)	211	43
Department of Social Security	DSS	18	29
Department of Trade and Industry	DTI	1 000	126
Department of Energy	Den	78	–
Northern Ireland Departments	NI departments	526	190
Scottish Executive	SE (formerly SO)	486	1 670
National Assembly Wales	NAW (formerly WO)	11	2
Other departments	Other departments	1 070	459
Total civil departments		*7 893*	*6 237*
Total civil R&D		*20 927*	*17 354*
Ministry of Defence		15 282	11 843
Total		**36 209**	**29 197**

Source: Department of Trade and Industry (2000b).

3. Private sector R&D

Research in industry and commerce or business takes place within a broad range of firms in all sectors, although there is a high degree of concentration of expenditure and activity in just a few areas. Industrial R&D is heavily concentrated in pharmaceutical companies, with the top three funders of R&D all being pharmaceutical companies (two of which have recently merged to form Glaxo

SmithKline, now the world's largest pharmaceutical company). These three firms account for 26% of all UK business R&D and the top 20 companies account for 67%. In 1998, a government consultative document noted:

"The amount of R&D undertaken by UK firms has declined relative to other G7 countries. Only in a few industries do UK companies spend comparable (or larger) amounts on R&D as a proportion of sales than other G7 countries. Falls in the R&D intensities of sectors such as metal products, machinery and equipment, and electrical apparatus have not been offset by increases in sectors such as pharmaceuticals. Expenditure on R&D by SMEs is generally low but those in high-technology sectors may have very high R&D-to-sales ratios." (HM Treasury and Department of Trade and Industry, 1998)

3.1. Growth in R&D spending by major area

Table 7 shows the broad sectors of the economy and R&D spending over the last decade in real terms. The data was generated by the Department of Trade and Industry for its *Research and Development Scoreboard.*[2] The data confirm the importance of the pharmaceutical sector whose research and development investment has doubled within a decade. The data also underline the recent importance of transport equipment which has increased significantly in real terms since 1995.

Table 7. **R&D spending by major area, 1987-98**

GBP million

	1987	1988	1989	1990	1991	1992	1993	1994	1995	1996	1997	1998
In cash terms												
Manufacturing: total	4 748	5 252	5 773	6 362	6 118	6 305	6 741	6 848	6 917	7 035	7 360	7 872
Chemicals	545	642	657	722	707	720	721	689	701	627	680	688
Pharmaceuticals	684	843	946	1 206	1 199	1 446	1 679	1 820	1 813	1 852	2 151	2 238
Mechanical engineering	432	424	635	532	538	580	665	761	683	668	709	730
Electrical machinery	1 234	1 441	1 420	1 566	1 329	1 258	1 386	1 218	1 245	1 313	1 181	1 320
Transport equipment	507	536	576	620	638	670	717	710	833	977	966	983
Aerospace	746	725	818	984	1 005	898	782	860	886	812	893	1 039
Other manufacturing	600	641	721	732	702	733	791	790	755	787	779	874
Services: total	1 587	1 670	1 877	1 956	2 017	2 184	2 328	2 356	2 337	2 396	2 297	2 359
Total	**6 335**	**6 922**	**7 650**	**8 318**	**8 135**	**8 489**	**9 069**	**9 204**	**9 254**	**9 431**	**9 657**	**10 231**
In real terms (base year = 1998)												
Manufacturing: total	7 545	7 817	8 019	8 195	7 420	7 402	7 709	7 721	7 580	7 469	7 600	7 872
Chemicals	866	956	913	930	857	845	825	777	768	666	702	688
Pharmaceuticals	1 087	1 255	1 314	1 553	1 454	1 698	1 920	2 052	1 987	1 966	2 221	2 238
Mechanical engineering	687	631	882	685	653	681	761	858	748	709	732	730
Electrical machinery	1 961	2 145	1 972	2 017	1 612	1 477	1 585	1 373	1 364	1 394	1 220	1 320
Transport equipment	806	798	800	799	774	787	820	801	913	1 037	998	983
Aerospace	1 186	1 079	1 136	1 267	1 219	1 054	894	970	971	862	922	1 039
Other manufacturing	954	954	1 001	943	851	861	905	891	827	836	804	874
Services: total	2 522	2 486	2 607	2 519	2 446	2 564	2 662	2 656	2 561	2 544	2 372	2 359
Total	**10 068**	**10 302**	**10 626**	**10 714**	**9 867**	**9 966**	**10 372**	**10 378**	**10 141**	**10 013**	**9 972**	**10 231**

Source: Department of Trade and Industry (1999b).

Table 8 indicates that for manufacturing industry, the amount of R&D carried out in the firms which have fewer than 100 employees is low. On average, this amount is around 3% of the total R&D carried out within all firms in manufacturing. In aerospace, the amount of R&D which takes place in the very small firm sector is very low, at around 1% of the total R&D occurring within the sector. By contrast, in non-manufacturing industry, the proportion of R&D which takes place in the smaller firms is much higher at around 14%.

Table 8. Business enterprise R&D: effects of sector and size, 1997

GBP million

Employment band	\multicolumn{8}{c}{Business enterprise R&D expenditure in 1997 analysed by sector and number of employees}
	0 to 99	100 to 399	400 to 999	1 000 to 4 999	5 000 to 9 999	10 000 to 19 999	20 000 and over	Total
Sector								
Chemicals	20	93	139	436	-	-	-	688
Pharmaceuticals	67	383	163	1 607	20	-	-	2 240
Mechanical engineering	48	142	165	294	82	-	-	730
Electrical machinery	37	188	404	636	41	-	14	1 320
Transport equipment	31	129	191	125	-	-	-	983
Aerospace	1	11	53	724	-	250	-	1 039
Other manufactured	55	254	182	290	91	-	2	874
Total manufacturing	259	1 199	1 295	4 110	-	250	-	7 872
Non-manufacturing	320	379	396	667	-	54	-	2 359
Total	**579**	**1 577**	**1 692**	**4776**	**629**	**304**	**674**	**10 231**

Source: Department of Trade and Industry (2000b).

Table 9 expresses R&D expenditure as a percentage of value added, comparing 1997 data with 1990 and the United Kingdom with the EU-9 and OECD-14. It shows a slight decline in the UK position over the period and, for aggregated classes, only in the medium-low-technology industries does UK industry exceed either average. However, the very high relative R&D intensity of the UK pharmaceutical sector is notable. Chemicals and petroleum refining also compare well internationally, while the aerospace sector is well behind.

3.2.   International comparison of R&D as a percentage of value-added

Table 9. Business enterprise R&D expenditure as a percentage of value added

Area	UK 1990	UK 1997	EU-9 1990	EU-9 1995	OECD-14 1990	OECD-14 1995
Total manufacturing	6.1	5.4	5.2	5.3	6.8	6.7
Food, beverages and tobacco	1.3	0.9	0.9	0.8	1.1	1.1
Textiles	0.3	0.4	0.3	0.5	0.6	0.8
Wood and wood products	0.2	0.1	0.3	0.6	0.5	0.6
Paper and printing	0.3	0.2	0.4	0.5	0.7	0.9
Chemicals	11.8	12.7	7.8	7.0	8.9	8.4
Industrial chemicals	7.8	6.3	8.5	7.4	9.1	7.7
Pharmaceuticals	34.5	32.3	25.0	23.1	22.6	22.4
Petroleum refining	16.5	20.4	3.0	2.0	5.2	3.6
Rubber and plastics	0.9	0.9	2.0	1.9	3.0	2.8

Table 9. **Business enterprise R&D expenditure as a percentage of value added** *(cont'd)*

Area	UK 1990	UK 1997	EU-9 1990	EU-9 1995	OECD-14 1990	OECD-14 1995
Non-metallic mineral products	1.3	1.0	1.0	1.1	2.2	1.9
Basic metals	1.7	1.0	1.6	1.4	2.6	2.3
Ferrous metals	1.4	1.0	1.4	1.5	2.3	1.9
Non-ferrous metals	2.5	1.1	1.9	1.3	3.4	3.0
Fabricated metals and machinery	9.1	7.4	8.8	9.2	11.5	11.5
Fabricated metals	0.8	1.1	1.2	1.1	1.3	1.2
Non-electrical machinery	4.5	3.9	4.9	5.9	4.7	5.6
Computers, office machinery	19.1	4.8	16.1	15.1	30.6	25.3
Electrical machinery	10.5	7.1	7.5	7.3	9.8	8.9
Communications equipment and semiconductors	16.2	13.7	19.4	18.6	16.5	17.2
Shipbuilding	3.0	1.2	2.5	4.7	1.4	2.2
Motor vehicles	8.9	10.8	9.2	10.6	12.7	12.7
Aerospace	19.4	18.0	33.7	35.3	37.2	39.4
Other transportation	3.9	4.9	3.7	7.1	7.2	7.0
Scientific instruments	4.3	3.2	4.5	10.8	11.4	19.5
Other manufacturing	1.7	1.5	1.3	1.7	1.9	1.9
High-technology industries	21.7	19.9	22.6	21.9	24.1	22.7
Medium-high-technology industries	7.2	6.4	7.2	8.0	8.9	9.4
Medium-low-technology industries	2.6	2.2	1.6	1.5	2.4	2.1
Low-technology industries	0.7	0.5	0.5	0.6	0.8	0.9

Source: OECD (2000).

3.2. *Types of R&D*

Table 10 shows the *Frascati* classification of business R&D activity in the United Kingdom for 1997 (using the *Frascati* three-fold classification). In terms of the types of research carried out by the private sector in the United Kingdom, the vast majority is of the applied type, with only a small proportion of basic research being undertaken. In 1997, only 5% of the R&D carried out privately in the United Kingdom was at the basic level. The only sectors in which a significant proportion of basic research takes place are Research and development services (30%) and Office machinery and computers (26%). In both cases, public sector contracts are likely to be a factor. R&D services is, in effect, an intermediary sector but its growth in the United Kingdom reflects a trend towards outsourcing of industrial R&D. Howells (1999) notes that the proportion of contracted out R&D relative to BERD in the United Kingdom grew from 5.5% in 1985 to 10.0% in 1995. He argues that the contract research and technology market has become a crucial part of the innovation infrastructure of the national economy.

Table 10. **Business R&D by *Frascati* category (3-category system), 1998**

Category	%
Basic pure	4.8
Applied	38.8
Experimental development	56.4
Total	**100.0**

Source: Department of Trade and Industry (2000b).

4. Operation of industry-science relationships in the United Kingdom

4.1. Typologies of interaction

There is a large and detailed literature on the topic of linkages between industry and the science base. A recent annotated bibliography from the British Library provides a guide to some of the newer material with a UK slant (Stewart, 1999). Among the most influential writings in recent years is the volume by Gibbons *et al.* (1994) which argued that there has been a transformation from traditional knowledge, called "Mode 1", generated within a disciplinary, primarily cognitive context, to "Mode 2" knowledge created in broader transdisciplinary social and economic contexts and carried out primarily in the context of application. This much-cited work has very clear implications for how we look at the issue of linkages:

> "...the notion of technology transfer has to be reconsidered. It cannot any more be understood as a transmission of knowledge from the university to the receiver easily and usually with almost no follow up. Instead it is no longer like a relay race, in which the baton is passed cleanly and quickly from one runner to the next. Technology transfer looks more like a soccer game in which the university is a member of a team. To score it needs the aid of all its team mates. The ball is passed back and forth among the players who may include businesspeople, venture capitalists, patent attorneys, production engineers, and many others in addition to the university faculty. This is why it has been suggested that technology interchange is a more appropriate phrase than technology transfer." (Gibbons *et al.*, 1994, p. 87)

A growing understanding of the interactive nature of science-industry links has been a feature of the best analyses in the literature for very many years. Table 11 summarises key findings. In the first study of its kind, by Carter and Williams as long ago as 1957, the authors were already detecting that scientific publication was a medium which would not reach the great majority of firms. They stressed the importance of communication networks, informal contacts and what would today be called intermediaries. The next major milestone in unravelling the linear model was the work by Langrish *et al.* (1972), which, on the basis of extensive case studies, found that scientific breakthroughs leading to radical new products or processes were the exception rather than the rule, with industry laying much greater stress on provision by universities of techniques and trained people. Building on this work was a framework developed by Gibbons and Johnston (1974) and extensively reworked in an up-to-date context by Faulkner and Senker (1995).

To conclude this very short review, we refer to an influential study carried out by SPRU for HM Treasury, which examined *The Relationship between Publicly Funded Basic Research and Economic Performance* (Martin and Salter, 1996). Itself a literature review, this report concluded that there were six different types of benefit from basic research, among which only a comparatively small proportion of benefits flow in the form of new useful knowledge that is directly incorporated in new products or processes.

For the purposes of this report, the typology used is a simple one, that which structures the United Kingdom's surveys on industry-academic linkages:

- Research contracts, collaboration and consultancy.

- Commercialisation of research.

- Teaching and training.

The first two of these are reported in this section and the third in Section 6.

Table 11. **Key references on UK science-industry linkages**

Reference	Sources and modes of interaction
Carter, C.F. and B.R. Williams (1957), *Industry and Technical Progress*, Oxford University Press.	• Scientific papers and conferences • General scientific and technical press • Trade journals • Communication networks and informal contacts with industrial researchers • Research associations and other scientific middlemen
Langrish, J., M. Gibbons, W.G. Evans and F.R. Jevons (1972), *Wealth from Knowledge – A Study of Innovation in Industry*, MacMillan, London.	• Provision of techniques of investigation to solve industrial problems • People trained in techniques and scientific thought • Occasional scientific breakthroughs
Faulkner, W. and J. Senker (1995), *Knowledge Frontiers – Public Sector Research and Industrial Innovation in Biotechnology, Engineering Ceramics, and Parallel Computing*, Clarendon Press, Oxford, building on M. Gibbons and R. Johnston (1974), "The Roles of Science in Technological Innovation", *Research Policy* 3, pp. 220-242.	• Knowledge of particular fields – scientific theory, engineering principles. Properties, etc., knowledge of knowledge • Technical information – specifications and operating performance of products or components • Skills – specific skills such as programming, hardware design • Knowledge related to artefacts – process or research instrumentation, other intermediaries (reagents, etc.)
Martin, B. and A. Salter (1996), *The Relationship Between Publicly Funded Basic Research and Economic Performance – A SPRU Review*, Science Policy Research Unit, Report prepared for HM Treasury, July.	• Increasing the stock of useful information • New instrumentation and methodologies • Skilled graduates • Professional networks • Technological problem-solving • Creation of new firms

Source: Cox *et al.* (2000).

4.2. Research contracts, collaboration and consultancy

An analysis of the level of industrial and government funding of research in the UK higher education system shows that overall, over the period from 1995-98, the level of industrial and public corporations' funding increased both in real terms and as a proportion of the total funding for research activities carried out in the sector. Table 12 shows how this level has increased.

Table 12. **UK HE funding: industry, commerce and public corporations' share of total**

GBP thousand

	1994-95	1995-96	1996-97	1997-98
Total funding for research[1]	1 439 982	1 544 350	1 633 993	1 723 998
Industrial funding for research	157 227	169 137	187 703	206 599
Industrial funding as a % of total research	10.9	11.0	11.5	12.0

1. Does not include HEFCE contribution.
Source: HESA, Holis Database (2000).

The Higher Education On Line Enquiry database (HOLIS) can be used to provide an analysis of funding patterns within different subject areas. Table 13 gives the levels of industrial and government funding of research in the higher education sector and shows clear changes over the period. The table ranks the different subject areas (actually "cost centres" in the HESA system) by the proportion of industrial research income in total research income in 1997-98 as an indicator of the orientation towards industry.

The rate of growth is calculated by subtracting the average of the final two years from the average of the first two years. For significantly sized areas, at the top of the list showing the most growth in the period are Pharmacology and Electrical, electronic and computer engineering. However, growth in the second of these areas is likely to be the result of new reporting and re-categorisation of expenditure by the Higher Education Statistics Agency.

In such subject areas as Architecture, Built environment and planning, Health and community studies, Clinical dentistry, Business and management studies, Librarianship, communication and media studies, and Agriculture and forestry, the trend is for industrial and public corporation funding to decline, with Agriculture and forestry showing the most significant decline over the period.

Perhaps surprising in this table is the position of Biosciences and some of the medical subject areas that currently show stable funding from industry. Further investigation of the position of the amounts of funding within the subject areas shows that industrial and public corporation funding is not correlated with total funding.

Figure 3 shows that the distribution of industrial income across universities is highly skewed. In fact, the top ten universities in terms of industrial income account for 43% of total industrial research income.

Figure 3. **Distribution of research grants and contracts income from UK industry**

UK universities, 1997-98, GBP thousand

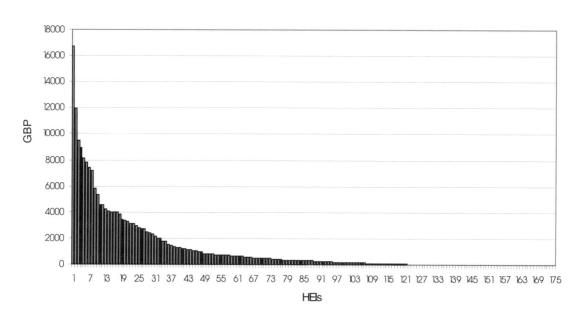

Source: PREST, compiled from HESA statistics.

Table 13. **UK HE sector: industry, commerce and public corporation research funding as a share of total research income, 1995-98**

Cost centre	Industrial funding as a share of total research income				Growth	Total income (GBP thousand) 1994-95 to 1997-98		Average yearly industrial income
	1994-95	1995-96	1996-97	1997-98		From all sources	From industry	
01 Clinical medicine	9.0	8.7	9.3	9.9	0.8	1 662 688	169 011	42 253
21 Mechanical, aero and production engineering	30.2	30.4	29.2	31.1	-0.1	201 988	87 453	21 863
10 Biosciences	6.2	5.4	6.4	7.0	0.9	789 358	53 327	13 332
11 Chemistry	14.8	14.7	16.7	17.5	2.4	271 953	51 782	12 945
20 Electrical, electronic and computer engineering	11.9	12.6	14.0	16.5	3.0	263 276	43 008	10 752
27 Business and management studies	30.5	29.6	29.8	26.7	-1.8	83 205	34 043	8 511
16 General engineering	19.6	20.0	21.2	24.2	2.9	116 160	31 733	7 933
18 Mineral, metallurgy and materials engineering	20.0	24.3	23.4	24.5	1.8	97 923	29 364	7 341
14 Earth, marine and environmental sciences	16.6	14.4	15.3	15.5	-0.1	153 614	27 997	6 999
13 Agriculture and forestry	18.2	21.0	15.8	14.8	-4.3	109 362	22 674	5 669
12 Physics	4.1	4.0	5.1	4.7	0.8	402 947	18 912	4 728
19 Civil engineering	15.5	17.8	17.7	17.6	1.0	79 694	16 490	4 123
25 Information technology and systems sciences	8.1	12.5	12.9	10.9	1.6	117 215	14 604	3 651
08 Pharmacy	26.8	24.6	26.6	29.1	2.1	39 426	14 505	3 626
29 Social studies	3.8	4.1	3.6	4.9	0.3	254 338	10 909	2 727
28 Geography	14.3	15.3	17.3	12.4	0.0	55 526	9 670	2 418
03 Veterinary science	8.9	11.3	10.8	12.3	1.5	73 219	8 902	2 226
09 Pharmacology	9.7	9.3	13.7	13.5	4.0	65 268	8 716	2 179
17 Chemical engineering	9.5	9.4	10.1	14.1	2.7	69 153	8 413	2 103
07 Psychology and behavioural sciences	6.9	8.1	7.8	7.9	0.4	98 376	8 191	2 048
02 Clinical dentistry	19.9	18.0	18.4	18.4	-0.5	33 583	7 704	1 926
04 Anatomy and physiology	4.5	3.9	3.8	4.3	-0.1	146 554	6 351	1 588
32 Humanities	7.3	7.9	9.7	11.4	2.9	58 671	5 967	1 492
23 Architecture, built environment and planning	11.7	10.0	8.4	12.9	-0.1	46 719	5 677	1 419
24 Mathematics	6.2	5.4	5.2	6.4	0.0	77 653	4 819	1 205
22 Other technologies	19.7	15.9	15.7	30.6	5.4	22 593	4 796	1 199
33 Design and creative arts	7.0	15.5	14.7	11.2	1.7	15 993	2 203	551
15 General sciences	18.6	13.1	14.8	20.6	1.8	9 378	1 782	446
06 Health and community studies	2.1	5.8	3.8	3.2	-0.5	43 089	1 666	416
31 Language-based studies	3.1	4.3	7.4	1.8	0.9	26 477	1 206	302
05 Nursing and paramedical studies	2.2	2.7	3.6	5.3	2.0	27 174	1 067	267
30 Librarianship, communication and media studies	10.1	13.3	17.8	1.9	-1.9	7 856	1 032	258
26 Catering and hospitality management	11.7	25.0	21.7	21.4	3.2	2 531	639	160

Source: PREST, based on HESA data (2000).

Figure 4. **Distribution of research grants and contract income by type of industrial collaboration, 1996-97**

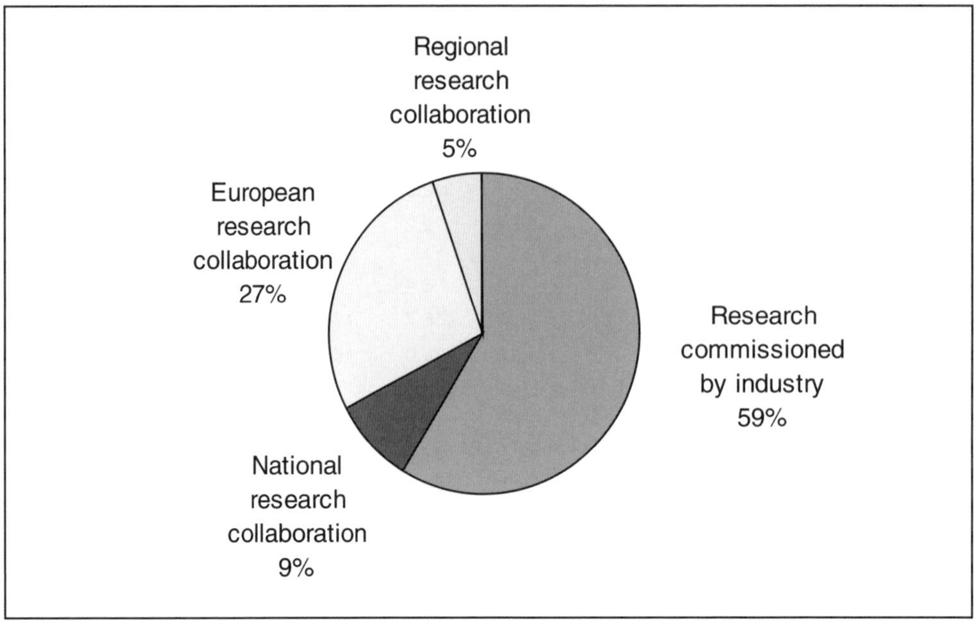

Regional
research
collaboration
5%

European
research
collaboration
27%

Research
commissioned
by industry
59%

National
research
collaboration
9%

Source: Howells *et al.* (1998).

Industrial funding for higher education institutions (HEIs) may be contracted directly or occur within the context of a public initiative. The PREST survey of university industrial liaison officers examined sources of funding which were linked to industrial research collaboration, including income from government attached to collaborative research with industry. It was found that just under 60% is commissioned directly by industry, and a quarter (27%) is via European-supported research collaboration with industry (Figure 4). A further 9% and 5%, respectively, is associated with UK collaborative research programmes such as LINK (see next section), and with collaborative research undertaken in the context of regional collaborative arrangements.

4.3. Commercialisation

Licensing and dissemination activities

The licensing of the intellectual property generated within universities is a relatively small area of ISRs but such data as is available (Howells *et al.*, 1998) suggests that the rate of growth in this area may be high. The production of intellectual property which ultimately finds industrial application can take a number of forms. At one extreme, intellectual property may be given free to industrial users with the proviso that there is a right to publish the findings of the research. At the other extreme, research can be carried out by the university but patented by the industrial funder. A combination of these types involves some collaboration in the carrying out of the work and/or sharing of the resulting intellectual property between the university and the funding firm.

For the two years for which data exist for the UK higher education system, there was an increase in the number of licences, which grew from 139 in 1995-96 to 177 in 1996-97. Regional differences were visible in the case of the United Kingdom where income from licensing rose in England and Scotland and fell in Wales. Overall, income from licensing appears to be increasing but the overall level, at under 1% of all research income and under 10% of all research income from industry, shows that the activity remains of limited importance within the current industry-science relationship.

Creation of spin-off firms

The United Kingdom has a growing number of spin-off businesses that have been set up by universities to commercialise a particular research potential. In 1998, a study found that around half of the surveyed universities had set up wholly or partially owned companies to exploit research (Howells, *et al.*, 1998). A total of 223 companies were identified, including holding companies for other spin-offs and intellectual property. Four of the responding institutions reported that companies which they owned had gone into liquidation during the period 1995-97, confirming the impression that spin-off businesses are usually open to considerable business risk. Of the firms operating in science parks that are university-owned, the majority are working in the biotechnology, life sciences and medicine, with engineering in second place.

Table 14, which is reproduced from the PREST report (Howells *et al.*, 1998), classifies the university-owned firms by subject area.

Table 14. **HEI-owned start-ups by field of science**

Percentages

Field of activity	%
Engineering	20
Biotechnology	19
Software	11
Chemical physical	11
General consultancy	10
Life sciences	9
Medicine	5
Other	15
Total	100

Source: Howells *et al.* (1998).

Operation and support for spin-offs: incubators, science parks

Universities (and to a lesser extent, public laboratories) have been closely linked with the emergence and development of science parks in the United Kingdom. Policy initiatives have largely centred on property-led schemes around science parks and related research and technology parks. Goals for science parks include: earning revenue; capturing more satisfactorily IPR leaking out of the university; attracting companies who may then become customers for the universities' research; and fulfilling a wider economic regeneration role within the local economy. However, other science parks have few or no ties with universities. The growth in the number of firms on university science parks within the United Kingdom (UKSPA, 2000) suggests that a period of rapid growth in the number of businesses operating in UK science parks has come to an end. Whereas the number of tenant companies on UK science parks rose from 301 firms to 1 020 in the period 1985-91, representing a rate of increase of 240% over the period, in 1991-97, it increased from 1 020 to 1 414, an increase of just 40%. While the latter half of the period still shows a high rate of growth, the trend is certainly very much reduced. It should also be noted that many of the firms established on science parks are not formally connected to the local university or indeed to any university. UKSPA data indicate that only 3% of firms based on science parks are HEI-owned. Moreover, a number of research studies have shown that the proportion of firms operating on science parks that are started by universities is relatively low – less than one-quarter, according to Massey *et al.* (1992).

Recently, the science park concept has itself given birth to so-called "incubators". Incubators are a supporting environment usually consisting of office premises, rather than whole buildings as in a science park. These are provided within existing or new buildings in close association with a university. Incubators also include some on-site management expertise and are normally focused upon a particular technology or subject area, such as biotechnology. Incubators are therefore smaller in scale than science parks but they are intended to provide much of the infrastructure in which academic-industry interactions can occur. The success of incubators is, as yet, not extensively researched.

Joint research centres and strategic partnerships

The United Kingdom has a long history of joint R&D projects and establishments. These joint establishments were set up to allow researchers working in basic and applied research to work on common problems and thereby to share tacit knowledge and develop novel techniques. This format is also favoured by overseas multinationals which have established laboratories in or close to UK universities (for example, the Hitachi Research Laboratory at Cambridge University).

In the pharmaceutical area, some of the larger firms are very strongly involved with large numbers of universities from the United Kingdom and abroad. Glaxo Wellcome (pre-merger) had a number of links which it terms "strategic partnerships" with universities around the world. There are currently four major joint developments in which the company is linking with specialist academic expertise in the United Kingdom and Wales. As the firm has historic ties to the United Kingdom, the relationship with the United Kingdom is strong. However, there are increasing links with universities from around the world, including one university in each of the following countries: South Africa, Canada, Singapore, France, Sweden, Australia, and six universities in the United States.

4.4.   *Industrial attitudes to collaboration*

The EU Innovation Survey available from Eurostat provides information on the intensities of ISRs in most of the nations of the European Union. Information about the United Kingdom is presented in a series of tables.

Table 15. **Sources of technological knowledge for innovation by type of innovator**

Sources of knowledge	Novel	Follower	Non-innovator
Commercial sources	97%	50%	24%
Knowledge pools	74%	27%	17%
Technology intermediaries	51%	31%	14%
Standards and regulations	51%	25%	9%
Science base	31%	5%	4%

Source: CIS2 data.

Using the CIS, UK firms have been broken down into three types: novel innovators, follower innovators and the innovatively-backward.

- *Novel innovators* introduced a technologically new or improved product between 1994-96 which was also new to their market.

- *Follower innovators* introduced a technologically new or improved product or process between 1994-96. This grouping excludes novel innovators.

- The *innovatively-backward* did not introduce any new products or processes between 1994-96.

Table 15 shows that the science base is the least important source of technological knowledge for innovation by comparison with other sources. Even among novel innovators, only 31% cited the science base and the numbers were far less for other types of firms.

Table 16 examines this topic on a sectoral basis, giving the percentage of firms in each sector which reported co-operation with universities for the UK and the EU average. Only in the most co-operative sector (energy, water supply) and in food and beverages is the United Kingdom substantially above the average. In most other sectors, the result is fairly close to average, although Computer and related services and Wood, paper, printing are lagging. Two important cautions should be borne in mind:

- There may be substantial variation in sub-sectors. For example in the last named category, Barker and Street (1998) have shown substantial variation in linkage patterns in the timber sector.

- Equally, there is substantial variation among firms within a sector.

Table 16. **Percentage of firms with co-operation with universities**

NACE / SIC	% UK	% EU
40-41 (energy, water supply)	36	20
23-24 (M: oil, chemistry)	25	21
30-33 (M: electrical, optical equipment)	18	17
15-16 (M: food, beverages)	16	9
34-35 (M: transport equipment)	16	15
27-28 (M: metals)	12	9
29 (M: machinery)	12	11
25-26 (M: rubber, plastic, mineral pr., glass)	11	8
74,2 (S: engineering)	10	10
72 (S: computer and related services)	9	12
17-19 (M: textiles, clothing, leather)	3	4
20-22 (M: wood, paper, printing)	1	6
36-37 (M: furniture, others)	1	2
65-67 (S: banks, insurance)	1	2
51 (S: wholesale trade)	1	4
64,2 (S: telecommunications)	0	8
60-62 (S: transport)	0	6

Source: European Commission (2000).

It is noticeable that very low levels of interaction are reported by major service sectors such as Banking and Wholesale trade. These sectors have benefited from innovations with science-base connections, although these may have been acquired through purchase of equipment and services from other sectors rather than in the context of a direct linkage.

5. Framework conditions for the United Kingdom's industry-science relationships

5.1. Introduction

This section of the UK country report examines and assesses the contexts that affect the way in which industry-science relations occur. These contexts are broad and include both legal and policy framework conditions.

5.2. Legal and employment contractual contexts for industry-science relations

In the United Kingdom, the legal frameworks in which the production, use and transfer of knowledge takes place have not exerted as profound an influence upon relations between industry and science as have policy initiatives. Some key developments are long-standing. Nevertheless, through laws and procedures relating to patents and employment contracts of employees, governments can exert some effect upon the industry-science relationship.

For the university sector up until 1985 a public body known as the National Research and Development Corporation (NRDC) had a monopoly in the exploitation of publicly funded research. This was ended by the then Conservative government with the intention that universities should take ownership of intellectual property generated as an incentive to engage with commercialisation (a similar logic to the US Bayh-Dole changes). The NRDC itself was restructured under the new name of the British Technology Group and subsequently privatised as an intellectual property management company. Universities formally had to satisfy the BTG that they had proper mechanisms in place for identifying, protecting and exploiting their intellectual property, and in particular for properly rewarding individuals.

Since the mid-1980s, many HEIs in the United Kingdom have developed far more sophisticated strategies for protecting and making use of innovations coming out of academic research. Many HEIs started to set up specialised intellectual property management and administrative centres, commonly known as technology licensing offices. These were set up within, or parallel to, existing industrial liaison offices.

For the public sector science research establishments, a recent report to the Treasury (Baker, 1999) recommended ways in which the commercialisation of their research could be increased. Baker's conclusions were gathered around four major points. First, the specific contexts of knowledge demanded that responsibility for commercialisation and use of research outputs should be local and devolved. This would require a fundamental change in the culture of organisations, with openness to the challenge presented by commercialisation. To this end, commercialisation of intellectual property would become a core goal of the laboratory. A second conclusion, which followed from the first, was that research organisations should be put at arms' length from the government, taking the form of the Scottish Office's agricultural research stations which have the status of non-department public bodies (NDPBs) but which are accountable to a Secretary of State. Such changes would entail new forms of employment contract for civil servants, and recommendations concerning personnel made up the third set of conclusions. The final set of conclusions asserted that in order to ensure effective commercialisation, new skills and new capabilities would be needed within the laboratories.

The report also observed that greater consideration should be given to the role of the industrial parties in the commercialisation process of public intellectual property. Baker suggested that future work towards commercialisation should focus far more upon "industry" push rather than science pull. Following the Baker Report, the government published draft guidelines (September 2000) on changes to the Civil Service Management Code (concerning the employment contracts of civil servants) allowing civil service employees to benefit from incentive-based pay linked to the commercialisation of research results. To help bridge the gap in finance for seed investments, the government will commit GBP 10 million (EUR 16.7 million) to a new fund for commercialising intellectual property, aimed at PSREs.

The following tables show how two research councils propose to reward employees for the commercialisation of intellectual property.

Table 17. **BBSRC: current incentive scheme, 2000**

Income from commercialisation (based on gross and net receipts)	Proportion of receipts paid to relevant staff involved in the commercialisation process
Gross receipts	
First GBP 1 000	100%
GBP 1 000 to GBP 50 000	20%
Net receipts	
GBP 50 000 to GBP 500 000	10%
GBP 500 000 to GBP 1 million	5%
Over GBP 1 million	2.5%

Source: OST (2000).

MRC: current incentive scheme, 2000

Income from commercialisation (based on gross and net receipts)	Proportion of receipts paid to relevant staff involved in the commercialisation process
GBP 5 000 to 14 000	100%
GBP 14 000 to GBP 80 000	33.3%
GBP 80 000 to GBP 600 000	25%
GBP 600 000 to GBP 1.5 million	20%
GBP 1.5 million to GBP 15 million	15%
Over GBP 15 million	10%

Source: OST (2000).

At the Defence Research and Evaluation Agency (DERA), changes to the way in which staff are employed have also been proposed. In the future, scientists who wish to commercialise their ideas will be deemed "DERA entrepreneurs". DERA entrepreneurs will work alongside DERA in the commercialisation of their own ideas, with DERA intending to take a share in the proceeds of their work. Such employment practices are new and untested, and it will take time for their effects to become evident.

In relation to employment practices, salary differentials between industry and science are widely thought to be preventing staff in industry from moving into an academic environment. Whether salary differentials are preventing the movement of staff from public laboratories to industry, and indeed whether such a movement would be beneficial are matters which it remains difficult to assess.

Financial regulation is thought by a number of groups to have an effect upon the performance of industry-science relations. The British Venture Capital Association (BVCA) argues that the current regulatory environment seriously impedes the creation of start-ups and spin-off firms because the current rate of capital gains tax at 25% is too high in comparison with other countries. The most usually cited comparator is the United States.

Box 1. Employment procedures for "DERA entrepreneurs"

Scientists wishing to exploit technologies arising in DERA through commercialisation in the civil sector are designated "DERA entrepreneurs".

A scientist wishing to become a DERA entrepreneur has to follow the modified business appointment rules laid down in DERA's Business Management System. In cases such as these, there are two variations to the standard business appointment application and procedure. They are:

- A series of additional considerations which must be addressed by the potential DERA entrepreneur and by the relevant line and functional department managers.

- A different clearance procedure. Under the procedure for business appointments, the approving authority is an individual personnel officer or other more senior manager (depending on grade).

However, all applications from DERA entrepreneurs are considered by the DERA Probity Board, which is established specifically to consider commercialisation projects.

The BVCA also argues in its budget submission to the government for 2001 that amendments to the Taxes Act 1988 should be made in order to allow entrepreneurs to benefit from the Enterprise Investment Scheme. The BVCA argues that such a change would not have an adverse effect upon the government's tax take. Other comments received suggest that there may be a need to extend tax credits for research and development more widely. Further comments from an intermediary organisation suggest that Clause 508 of the Income and Corporation Taxes Act (ICTA) 1988 should be amended to allow DTI flexibility in issuing corporation tax exemption certificates. It is argued that such a change would not be revenue neutral but would give rise to significantly higher tax revenues in the longer term because of the likely growth in the number of new business enterprises.

5.3. *Incentives and disincentives: Higher Education Innovation Fund (HEIF)*

There has been consistent criticism, particularly from industry, in the United Kingdom that the incentive schemes for universities and for individual academics do not sufficiently reward or motivate them for taking part in ISR activities. A focus of such criticism is the Research Assessment Exercise, which, as explained earlier, is used to allocate the bulk of block funding for research. Individual appointments and promotions are heavily influenced by the RAE and reflect traditional academic values. In an attempt to counterbalance this influence, a "third leg" of funding (*i.e.* in addition to teaching and research) was launched as the "Higher Education Reach Out to Business and the Community", known as HEROBC, although this initiative has now been subsumed within the Higher Education Innovation Fund (HEIF) which was launched after the government's spending review in 2000. HEIF receives funding from across government, from the Department of Trade and Industry/Office of Science and Technology, from the Higher Education Funding Council for England, and from the Department for Education and Employment (renamed the Department for Education and Skills in 2001). The broad funding base indicates a high level of support and commitment for third-leg activities across government. HEIF also fulfils the function of consolidating and simplifying what might be seen as a confusing array of third-mission support initiatives.

HEIF, and before it HEROBC, aims to develop the capability of HEIs to respond to the needs of business, and to contribute to economic growth and competitiveness, by enabling them to put into practice organisational and structural arrangements to achieve their strategic aims in this area. This fund is intended to initiate a permanent third stream of funding, complementing the Council's existing grant for teaching and research, to reward and encourage HEIs to enhance their interaction with business. HEIF provides a platform of core funding to help them to put into practice organisational and structural arrangements to develop and implement strategic approaches to their relations with business, and to assist in activity to improve the transfer of knowledge and skills.

This scheme has at its core the belief that all HEIs should be engaged with business in different ways. The fund is intended to enable them to develop links across the full range of their academic endeavours, and to develop closer working relations with business which could also influence the way in which they prepare their students for employment.

Under HEROBC two rounds of funding have occurred, the first being published in June 1999 and the second in February 2000. Future funding under HEROBC will be allocated on the basis of a formula which recognises the level of engagement of HEIs with business and the community. The formula is likely to encompass the main measures used in past surveys of linkages. Around GBP 140 million is likely to be allocated to HEIF over a three-year period, a step which will lead to a tripling of the funds available to third-leg activities in only a few years.

5.4.   *Policy contexts for industry-science relations*

In the United Kingdom, the policy context is of fundamental importance for industry-science relations. A whole range of relevant institutions – governments, their agencies, industrial sectors, higher education institutions and intermediaries – have all contributed to the development of a large number of initiatives intended to further the development of networks of collaboration. A wide variety of policy initiatives in innovation and technology policy, in education and labour policy and in fiscal and competition policy affect industry-science relations in some way.

The range of policies employed in the domains of research collaboration and commercialisation is shown in Table 19. It should also be noted that the desire to promote linkages pervades more general Research Council funding, with application forms seeking evidence of interest on the part of users of research. Many initiatives require matching funds from industry.

Innovation policy background

In innovation policy, the emphasis has been to promote industry-science links themselves as a means to achieve innovation, technological development and economic growth. The significant emphasis upon the creation of industry-academic links as a cornerstone of innovation and technology policy is evidenced by the government White Paper on Science and Technology (*Excellence and Opportunity, A Science and Innovation Policy for the 21st Century*). In its key proposals, the White Paper states:

"Government, with others, has a clear role in the funding of basic curiosity-driven research; the value of basic research can rarely be captured by the private sector. In addition, scientific knowledge is often produced through collaboration. Society benefits from basic knowledge being shared as widely as possible. The private sector will rarely invest in research when it cannot be confident of making a return.

"In addition there are often market failures in the networks and links which bring the public and private, researchers and industries together. Although all parties would benefit from these linkages being stronger, it is in no one party's interest to take on the cost and responsibility for

forming these networks. That is why public funding can play such a critical role."
(Cmnd. 4814, 2000)

Table 18. **ISR activities and policy support mechanisms**

ISR activity	Support mechanism

Joint research – Joint R&D projects, strategic partnerships and alliances	• Foresight Link
	• Link:
	    Electronics/Communications/IT
	    Food/Agriculture
	    Biosciences/Medical
	    Materials/Chemicals
	    Energy/Engineering
	• Joint Research Equipment Initiative (JREI)
	• IMI (Innovative Manufacturing Initiative)/ now LINK
	• Foresight Awards (Royal Academy of Engineering)
	• Faraday Partnerships
	• Higher Education Innovation Fund (HEIF) initially Higher Education Reach-Out To Business and the Community (HERO)
	• Space Technology Research Programme
	• Industrial Programme Support Scheme (PIPSS)
	• CONNECT (NERC only)
	• Joint Grant Scheme (through DERA)
	• Defence Aerospace and Systems Research Partnerships (DARPS)
	• Beacon (though DERA)
	• Industry Interest Groups (DERA SIGS)
	• Innovation Relay Centres (EU managed by DERA)
	• Biotechnology Mentoring and Incubator Challenge (BMI Challenge)
	• Realising Our Potential Awards (ROPA)
	• EUREKA

Licensing and commercialisation	• Patent Office Activities • Links with AURIL, Abolition of Patent Fees, Intermediaries Workshops • The Biotechnology Exploitation Platform Challenge (BEP Challenge) • Public Sector Research Exploitation Fund
Operation and support for spin-off companies – incubators and science parks	• University Challenge 1999 (Funded by the DTI, Wellcome Trust and Gatsby Charitable Foundation – eventually to be within HEIF) • Technology Ventures Scotland

Further key proposals of the White Paper underline the role of the industry-science relationship to innovation and the belief that government action to promote linkages will promote economic growth and development. The standard market failure justification for public support of science remains a core assumption for government policy in the United Kingdom.

Box 2. UK Government Science White Paper 2000: key points

Government can facilitate such links to help turn scientific ideas into innovation. This means examining new public-private partnerships to bring businesses and universities, ideas and finance closer together, as well as initiatives to create regional clusters. Government cannot and should not attempt to manage these networks but it can play a critical role in facilitating their creation.

Government must also provide the best framework for scientists and businesses to make international links. To play a full part in modern science and to bring its benefits to the United Kingdom we have to co-operate internationally. Our investment in science in the United Kingdom is an entry ticket to the global collaboration that is the driving force of scientific advice. Government plays a crucial role in making this possible. A society that is closed, inward looking and defensive would not remain at the forefront of science because it could not take part in this global collaboration. Britain is stronger when it collaborates internationally. Britain must be a key player in European and global science.

To extend opportunities for innovation the Government will:

- Establish a Higher Education Innovation Fund of GBP 140 million over three years incorporating the Higher Education Reach Out to Business and the Community fund to build on universities' potential as drivers of growth in the knowledge economy. This will triple existing funding by the third year, to increase universities' capabilities to work with industry, particularly small firms.

- Launch a new Foresight fund, initially up to GBP 15 million, to get the best ideas from Foresight 2000 put into action fast.

- Run one further round of the University Challenge Competition, to provide seed venture funding for knowledge transfer; double the number of new starts for Faraday Partnerships from four to eight a year, to link the science base to business networks; and put GBP 15 million more into Science Enterprise Centres to bring business skills into the science curriculum.

- Create new Regional Innovation Funds of GBP 50 million a year to enable Regional Development Agencies (RDAs) to support clusters and incubators and new clubs of scientists, entrepreneurs, managers and financiers.

- Support 20 Business Fellows who will lead their academic colleagues in working with business. They will spend part of their time advising companies, particularly SMEs, on their business problems, providing technical and research solutions.

- Publish science and innovation strategies for government departments.

- Introduce a Small Business Research Initiative to open up to small firms R&D procurement worth up to GBP 1 billion, with a target of procuring GBP 50 million of research from them.

- Change the rules for Government funded research, so that research bodies own the Intellectual Property Rights; issue new guidelines on incentives and risk-taking for staff in public service research establishments; and provide GBP 10 million to commercialise research done in the public sector, including the NHS.

- Double the number of International Technology Promoters from 8 to 16 and link their work closely with British Trade International and other UK agencies overseas, to help UK universities and businesses make new partnerships across the world. And we plan to extend the network of science attachés in embassies abroad.

Source: UK Government Science and Innovation White Paper, 2000, Cmnd. 4814.

Changes to the patenting process

Other innovation policies relevant to the industry-science interactions have originated from the Patent Office itself and concern aspects of the patenting process, although they do not involve specific legal changes. The changes proposed to the operation of the Patent Office concern the costs of patenting, and the provision of information about existing patents.

In order to encourage patenting, the Patent Office has undertaken to reduce charges for patenting or, in some cases, to remove them altogether. To increase general awareness of patents and new technologies, a new database based on the work of the Association of University Research and Innovation Links (AURIL) will be set up on the Patent Office Web site. In response to the recommendation of the Creative Industries Taskforce, the Patent Office created an intellectual property portal on the Internet that began to operate in late 2000.

Education policy

In educational policy, the emphasis has also been on the promotion of science and engineering to young people. The White Paper announced that the government would "make 2001/2002 Science Year and run a new Science Ambassadors programme to capture children's imagination and encourage them to take up careers in science and engineering."

Within educational policy, the creation of a new system of foundation degrees and a "university for industry" indicate a desire to involve the educational system in a closer relationship with industry. The University for Industry, or "Learndirect" as is has become known, will receive an GBP 84 million budget from the government for its first year of operation. Learndirect is the government's major lifelong learning initiative aimed at providing course material through 178 centres operated by locally based associations of "colleges, universities, local authorities, trades unions, companies and business organisations."

Foresight

Since 1994, the United Kingdom has been engaged in two rounds of a national Foresight Programme. The goals of the exercise are to:

- Develop visions of the future – looking at possible future needs, opportunities and threats and deciding what should be done now to make sure we are ready for these challenges;

- Build bridges between business, science and government, bringing together the knowledge and expertise of many people across all areas and activities; in order to

- Increase national wealth and quality of life.

The programme has operated through ten sectoral and three thematic panels. The latter have engaged problem-centred areas, dealing with the consequences of the ageing population, crime prevention and the future of manufacturing. Through an infrastructure of task forces and implementation activities, the programme is seeking to effect a cultural shift through the creation of new future-oriented networks. At the time of writing, each panel has reported and a synthesis report from the Programme Steering Group is in preparation. Following ministerial criticism, the Programme has been reviewed with the result that Foresight activities will have far stronger regional orientations in the future.

Among the follow-up measures to the first round of Foresight was a dedicated scheme, the Foresight Challenge Competition. This was launched at the end of 1995 with the explicit aim of increasing interaction between industry and academia. Consortia of business and the science base were able to apply for matching funds for projects addressing Foresight priorities. In the first round, following a large number of applications, awards were made to 24 projects costing a total of GBP 92 million, of which GBP 62 million came from industry and GBP 30 million from the OST. The second round of the initiative, renamed Foresight LINK Awards, makes available GBP 10 million of government funding. The SHEFC and HEFCW also provided funding for research projects reflecting Foresight priorities, awarding GBP 7.5 million and GBP 1 million, respectively, to support a total of 25 projects. Among the respondents to the questionnaires, nearly half commented that Foresight and LINK activities have been the most effective government mechanisms in the promotion of industry-science relations. A new Foresight fund, initially up to GBP 15 million, has been announced to follow up ideas emerging from the current programme.

LINK

The LINK scheme is the government's principal mechanism for promoting partnership in pre-competitive research between industry and the research base. It aims to stimulate innovation and wealth creation, and improve the quality of life. The scheme offers an opportunity to engage with some of the best and most creative minds in the country, to tackle new scientific and technological challenges so that industry can go on to develop innovative and commercially successful products, processes and services. LINK focuses on areas of strategic importance for the future of the national economy. All new programmes address priorities under the government's Foresight Programme. Companies and research organisations throughout the United Kingdom can participate in LINK projects. Small and medium-sized enterprises (SMEs) are particularly encouraged to get involved.

Multinationals can also participate, provided they have a significant manufacturing and research operation in the United Kingdom, and the benefits of research are exploited in the United Kingdom or European Economic Area. LINK covers a wide range of technology and product areas from food and bio-sciences, through engineering to electronics and communications. Each LINK programme focuses on a particular technology or market area. Overall programme goals are defined by the sponsors, in consultation with industry and the research base. Each programme supports a number of collaborative research projects. A typical project lasts between two and three years, and brings together research base and industrial partners, within a well-defined project management framework. Each project must contain at least one firm and one science-base partner. A collaboration agreement, drawn up by the partners, specifies how the fruits of the research will be shared. LINK programmes are sponsored by government departments and Research Councils. Programme goals are defined by the sponsors, in consultation with industry and the research base, taking account of priorities identified by the Foresight Programme. LINK stimulates interdisciplinary research in areas such as:

- Sensors.
- Medical engineering.
- Advanced food science.
- New communication systems.
- Future vehicles.
- Surface engineering.
- Catalysis.

Networking is strongly encouraged so that participants can share in the programme's achievements, supported by newsletters, seminars, technology transfer clubs and the LINK Web site.

Faraday Partnerships

The establishment of Faraday centres has had a long and difficult "birth". The centres, of which there are now 17 in number, were proposed as far back as 1992 by the Centre for the Exploitation of Science and Technology (CEST) as a means of bridging academia and SMEs in the United Kingdom. They were inspired by Germany's network of Fraunhofer Institutes. Under the previous Conservative government, the DTI was to provide matching funding with the EPSRC, but this did not happen and in the vacuum the EPSRC decided to go it alone. In September 1997, it therefore announced the establishment of four pilot Faraday Partnerships which would each receive GBP 50 000 in start-up funds and then up to GBP 1 million over four years from EPSRC. The main objectives of the Faraday Partnerships are to encourage greater interaction between HEIs and industry, especially SMEs. The aim is to expand information flows and links, thereby improving awareness in academia of industry requirements and increasing exploitation of research results.

Of the four initial Faraday centres, three involve Partnerships between universities and independent research and technology organisations. The centres are in the domains of enhanced packaging technology, 3D multimedia applications and technology integration, interdependent mechanical and electronic parts, and intelligent sensors for control technologies. One of the intermediary organisations interviewed for this study believed that Faraday was the most effective of all government programmes to promote industry-science links.

The DTI/EPSRC have announced that they will establish four new Partnerships in each of the financial years commencing 2000-01, so that by 2003 there will be a national network of 16 Partnerships in addition to the existing four, bringing the total to 20.

University Challenge Fund

The University Challenge Fund competition assists winning universities or consortia of universities in setting up local "seed" funds to support the early stages of commercialisation of academic research. Each fund is managed by a board, normally with venture capital expertise present. The funds may finance further research in support of commercialisation, the cost of patenting, building prototypes, market research and the preparation of business plans to attract next-stage capital. The fund was initially supported by the DTI, the Wellcome Trust and the Gatsby Charitable Foundation. In the first round, 15 awards to a value of GBP 25 million were made which, with matching funding from universities and other sources, created funds of GBP 60 million. A second round of funding plans to disburse in the second round of the Challenge Fund.

5.5.   *Incentives and barriers for scientific institutions*

Incentives and barriers for higher education institutions

For scientific institutions, the major incentives for collaboration and the formation of academic-industry links are firstly, and most importantly, access to funding (Table 20). In a period of increasing financial stringency in the higher education sector, the attempt to generate further funding from whatever source, including government, charities, industry and overseas, has been strong. Previous research also reveals that collaboration and the creation of links with industry is a strategic goal that many institutions have readily embraced.

A number of other incentives are also present for creating and maintaining industry links. The need to find a way to exploit existing research capabilities – showing an economic rent from the human capital of the organisation – is an incentive. The opportunities for universities in this area are thought to have improved in the last ten years as the size of the research base of the public research labs has been reduced. Government policy research has therefore been increasingly carried out within the academic sector. Access to equipment and other resources has also been given as a further incentive for higher education organisations particularly to form relationships with industrial companies. Access to relevant expertise within industry has also been given as a possible reason for participation by science in relationships with industry.

Table 20. Factors motivating links with industry in terms of research contracts and income ranked by mean value (all UK HEIs)

Rank	Motive	Mean
1	To access industrial funding	4.2
2	Collaboration with industry is a strategic institutional policy	2.6
3	To find an exploitation outlet for research capabilities	1.9
4	To access complementary expertise	1.6
5	To provide an outlet for research results	1.5
6	To access state-of-the-art equipment & facilities	0.9
7	To contribute to local economy	0.7
8	Government policy and/or political pressure	0.5
9	To contribute to UK economy	0.4

Source: Howells *et al.* (1998).

Of the barriers which make higher education institutions reluctant to create linkages with industry for the purposes of research or consultancy, the most important appear to be differences in objectives, with different sets of priorities often held by each party. Another major barrier which has been suggested by research carried out by PREST (Howells *et al.*, 1998) is that the nature of the work required by industry fails to provide the necessary stimulation which academics expect. Other barriers include locating suitable partners for the research. This problem of locating partners is often significant but could be lessened through wider use of information technology resources. Further barriers include restrictions on the publication of research results and their commercialisation where this becomes the chosen route. The movement of staff within the industrial partner organisation is also given as a possible reason why universities do not enter into collaborative arrangements with industry, may fail to remain in them and generally are unable to generate any substantial benefits from ISRs.

Table 21. Barriers to establishing research links with industry

Ranked by mean value (all UK HEIs)

Rank	Barrier	Mean
1	Differences in objectives	2.59
2	Work needed by industry not interesting	1.84
3	Getting in touch with relevant industrial organisations	1.81
4	No influence on base-line funding	1.56
5	Insufficient equipment & facilities	1.36
6	No influence on academic promotions	1.21
7	Delay in publications	1.18
8	IPR issues	1.14
9	HEIs not seen as reliable	0.99

Source: Howells *et al.* (1998).

Table 21 shows the major barriers cited by University Industry Liaison Officers (ILOs) in their assessment of the major obstacles to research links between industry and academe.

Incentives and barriers for firms

Individual firms are active in industry-science relations to the extent that they perceive the benefits from collaboration and exchange to be greater than the costs of participation.

Motivations for firms to enter into such linkages have not been studied systematically in the United Kingdom, but case-study work suggests that the following are important:

- Securing a supply of recruits.

- Access to scarce scientific expertise.

- Access to exploitable intellectual property.

- Enhancing the scope and testing of in-house corporate activities.

- Scanning and entry to "invisible college".

- Cost saving through outsourcing.

The barriers which firms believe prevent them from entering into effective and productive relationships with science institutions include the following:

- Lack of a professional approach by the institution.

- Divergence of objectives between the partners.

- Misunderstandings and lack of precise aims.

- Apparently low priorities given to the work by academics which downgrades the role of ISRs.

- Maintaining contacts with the inevitable turnover of employees within the industrial firm.

- High perceived costs of locating partners and participation.

Those representing industry also argue that the failure to agree on the terms of any collaborative agreement is affected by the priorities of the academic collaborator. As noted above, the Research Assessment Exercise (RAE) is cited by industrial commentators as a principal reason why academic staff avoid involvement in industry-science relations. The pressure of the RAE has been experienced across the whole range of institutions. Among the older, traditional institutions, ISRs are often seen as a distraction from the attempt to protect or increase the ratings scores of individual departments, ratings that are achieved largely through the quality of the publication records of their staff. Among the newer universities, which have been attempting to build their academic reputations following their conversion from polytechnic status, collaboration with industry has also been viewed as a distraction.

5.6. Role of intermediaries

The number of intermediaries in the area of industry science links has grown significantly over recent years. Intermediaries and/or consortia often provide an essential mechanism in cases where if they did not exist, industry-academic links would not occur. There are two major rationales for their existence:

- In scientific or technological fields, one firm often cannot afford to fund, or risk funding, the research.

- Sometimes, a single HEI cannot supply all the scientific and technical capabilities of the required research.

There are situations where an intermediary provides a link between a firm(s) and a suitable HEI(s), which otherwise would not occur through lack of information, or the high costs of information scanning for the firm (especially in relation to an SME). There are more complex areas where a single firm on its own would not fund the research and needs an impartial arbiter and "animator" to set up a project. This would occur in situations such as research on aspects of industry standards or use of shared facilities or networks.

Many of these intermediaries are local or regional in character. Good relations with these bodies and agencies have been seen as increasingly important, particularly when paralleled with HEIs' growing recognition of their local and regional responsibilities. In particular, links both directly and through their intermediary role with TECs, LECs, Chambers of Commerce, Business Links, new and existing national and regional development agencies, former development corporations, RTCs and local authorities have therefore gained increasing significance.

In the United Kingdom, the pharmaceutical and biotechnology industries have strong intermediary organisations whose impact on industry science relations has increased significantly within the last three years, according to interview evidence collected for this study. The two organisations – the Association of the British Pharmaceutical Industry (ABPI) and the BioIndustry Association (BIA) – have broad concerns, BIA in particular being concerned with support for SMEs which have been adversely affected by the decision of larger pharmaceutical firms to base their manufacturing processes outside the United Kingdom to reduce costs.

The Regional Development Agencies Act of 1998 gave new impetus to regional innovation policy, making each regional development agency responsible for furthering economic development, regeneration of its area, the promotion of business efficiency, investment and competitiveness in its area, employment, and the enhancement of skills relevant to employment in its area.

6. Organisation of industry-science relations in the field of human-capital production and exchange in the United Kingdom

6.1. Introduction

This section of the UK country report examines the organisation of industry-science relations in the context of the production of human capital. The report initially considers the contexts in which the production of graduates and postgraduates takes place, examining the flow of graduates and postgraduates in terms of their academic field to their sector of employment. The section then considers the substantial policy mechanisms operating in the United Kingdom for training and mobility of research personnel through the medium of industry-science relations.

6.2. *Employment of graduates*

Table 22 shows the employment of all graduates (including undergraduates, masters and doctorate-level students) by field of science and by sector of employment. Data for 1998-99 are presented. The largest academic field from which graduates are produced is that of the social sciences (around 60 000 students), including law and business, while the smallest category is that of engineering sciences (with around 15 000 graduates). Natural sciences is the second largest field of science from which graduates are produced, while the humanities is almost as large. Business and public services are the economic sectors to which the majority of graduates, including those qualified at postgraduate level, are recruited. For example, public services take 50.95% of all graduates from social sciences.

In terms of postgraduate training, the largest amount of postgraduate research and training at PhD level takes place in the natural sciences, where 2 015 PhDs were produced in 1998-99. Engineering sciences produce the second highest level of PhDs (around 546 in 1998-99), with medicine in third position, producing just over 500. Humanities and social science are the smallest of the fields of science in terms of their production of PhDs, each providing just over 300 PhDs in 1998-99 to employment.

In terms of the flow of postgraduates with PhDs into employment, Table 23 shows that the humanities contribute around 66% of their PhDs to education (which is the largest of all the fields of science), while engineering sciences contribute only around 34% of their PhDs. Those postgraduates with PhDs moving into research and development activities include mostly those with a natural sciences background (8.3% of all doctorates in natural science), closely followed by those from engineering sciences (7.1% of all doctorates in engineering sciences) and medical science (6.7% of all doctorates in medical science).

Table 24 shows the proportion of PhD graduates entering a particular field and indicates the extent to which a particular industrial sector is research intensive. This table covers those sectors in which the percentage of entrants with PhDs exceeds 1.5%. These sectors where the graduate recruitment includes a higher proportion of those with a research degree – PhD in this analysis – are the research and development area, the chemicals sector, the manufacture of basic metals, and the oil and gas exploration area. A large number of PhD graduates go into the education sector, although the proportion of PhD entrants is only 5.6% of all graduates entering the sector. Of all the 69 industrial sectors, only 27 have a recruitment rate for PhDs as a proportion of all graduates greater than 1%. The requirement needs of sectors for postgraduates with doctorates is therefore highly concentrated within a small number of sectors.

Table 22. **Employment of all graduates by field of science and industrial classification, 1998-99**

Industrial classification	Humanities		Social science		Natural sciences		Medical science		Engineering sciences	
	Total	%	Total	%	Total	%	Total	%	Total	%
Agriculture, forestry & fishing	56	0.23	92	0.15	177	0.64	6	0.03	531	3.48
Mining	33	0.13	108	0.18	224	0.81	1	0.01	89	0.59
Manufacturing										
Food & tobacco	138	0.56	503	0.84	300	1.07	31	0.16	297	1.95
Textiles & leather	283	1.14	104	0.17	108	0.38	4	0.03	22	0.15
Wood products	18	0.07	15	0.02	16	0.06	0	0	12	0.08
Pulp, paper & printing	1 724	6.96	899	1.49	430	1.55	38	0.19	130	0.85
Coke, refined petroleum prods. & nuclear fuel	18	0.07	59	0.1	67	0.24	0	0	85	0.56
Chemicals & chemical products	134	0.54	335	0.56	1096	3.93	269	1.35	211	1.39
Rubber & plastic products	28	0.11	75	0.12	55	0.2	3	0.02	51	0.33
Other non-met. mineral prods.	90	0.36	46	0.08	54	0.19	0	0	27	0.18
Basic metals	13	0.05	42	0.07	71	0.25	1	0.01	89	0.58
Fabricated metal products, except machinery & equipment	40	0.16	67	0.11	58	0.21	2	0.01	103	0.68
Machinery & equipment nec	56	0.23	157	0.26	114	0.41	1	0.01	333	2.19
Electrical machin., computers	142	0.58	462	0.77	574	2.06	12	0.06	745	4.89
Medical, precision & optical instruments, watches & clocks	28	0.11	53	0.09	86	0.31	6	0.03	51	0.33
Motor vehicles, trailers & semi-trailers	59	0.24	181	0.3	97	0.35	2	0.01	382	2.51
Other transport equipment	25	0.1	114	0.19	128	0.46	2	0.01	409	2.68
Furniture; other manufacturing, recycling	221	0.89	124	0.2	118	0.42	1	0.01	89	0.59
Electricity, gas, steam & hot water supply, water	198	0.8	493	0.82	397	1.42	25	0.13	267	1.75
Construction	192	0.78	380	0.63	320	1.15	10	0.05	1 367	8.97
Wholesale, retail, motor vehicles, fuel, & hotels	4 365	17.62	5 885	9.8	2 956	10.59	1 206	6.06	930	6.1
Transport	569	2.31	1 240	2.07	495	1.79	39	0.2	233	1.53
Post and telecoms.	591	2.39	1 110	1.85	968	3.47	43	0.22	528	3.47
Insurance & financial	1 551	6.26	4 467	7.44	2 294	8.23	127	0.65	459	3.02
Business services	4 833	19.53	9 403	15.65	8 323	29.83	462	2.33	4 579	30.06
Public services	4 873	19.68	30 608	50.95	6 400	22.92	17 375	87.22	2 432	15.97
Sewage, sanitation & other services	3 974	16.04	2 402	4.01	1 486	5.33	169	0.86	506	3.32
Private households	47	0.19	49	0.08	30	0.11	4	0.02	25	0.16
International organisations	53	0.21	69	0.11	42	0.15	7	0.04	11	0.07
Unknown	432	1.74	551	0.92	440	1.58	74	0.37	252	1.65
Total	24 766	100	60 078	100	27 908	100	19 920	100	15 233	100

Source: HESA, data for PREST (2000).

Table 23. UK recruitment of PhDs in selected sectors and by field of science, 1998-99

Summary table

Sector	Humanities Total	Humanities %	Social science Total	Social science %	Natural sciences Total	Natural sciences %	Medical science Total	Medical science %	Engineering sciences Total	Engineering sciences %	All fields Total	All fields %
Education	209	66.6	204	61.6	805	40.0	213	42.0	188	34.4	1619	43.6
Health	11	3.5	23	6.9	188	9.3	133	26.2	15	2.7	370	10.0
Other business activities	13	4.1	26	7.9	133	6.6	14	2.8	70	12.8	256	6.9
Research & development activities	2	0.6	4	1.2	167	8.3	34	6.7	39	7.1	246	6.6
Computer & related activities	1	0.3	5	1.5	119	5.9	4	0.8	36	6.6	165	4.4
Public administration & defence; social security	16	5.1	28	8.5	72	3.6	3	0.6	24	4.4	143	3.9
Financial activities, except insurance & pension funding	4	1.3	10	3.0	39	1.9	2	0.4	5	0.9	60	1.6
Recreational, cultural & sporting activities	17	5.4	6	1.8	23	1.1	4	0.8	3	0.5	53	1.4
Publishing, printing & reproduction of recorded media	11	3.5	2	0.6	23	1.1	4	0.8	3	0.5	43	1.2
Manufacture of radio, television, communication equipment & apparatus	0	0.0	0	0.0	22	1.1	0	0.0	16	2.9	38	1.0
Manufacturing of food products & beverages	0	0.0	0	0.0	18	0.9	3	0.6	8	1.5	29	0.8
Extraction of crude petroleum & natural gas; service activities incidental to oil & gas extraction	0	0.0	0	0.0	26	1.3	0	0.0	1	0.2	27	0.7
Insurance & pension funding, except social security	1	0.3	1	0.3	14	0.7	0	0.0	1	0.2	17	0.5
Others	29	9.2	22	6.6	366	18.2	93	18.3	137	25.1	647	17.4
Totals	314	100.0	331	100.0	2 015	100.0	507	100	546	100	3713	100

Source: HESA, data for PREST (2000).

Table 24. **Recruitment of graduates by destination and proportion with PhDs, 1998-99**

Industrial sector	% of entrants with PhD
{73} Research and development activities	23.3
{24} Manufacture of chemicals and chemical products	13.5
{27} Manufacture of basic metals	9.3
{11} Extraction of crude petroleum and natural gas; service activities incidental to oil and gas extraction	7.3
{33} Manufacture of medical, precision and optical instruments, watches and clocks	7.1
{80} Education	5.6
{23} Manufacture of coke, refined petroleum products and nuclear fuel	5.2
{14} Other mining and quarrying	4.7
{41} Collection, purification and distribution of water (excluding sewage treatment)	4.6
{35} Manufacture of other transport equipment	3.7
{32} Manufacture of radio, television and communication equipment and apparatus	3.4
{31} Manufacture of electrical machinery and apparatus not elsewhere classified	3.0
{26} Manufacture of other non-metallic mineral products	2.8
{29} Manufacture of machinery and equipment not elsewhere classified	2.7
{72} Computer and related activities	2.7
{30} Manufacture and assembly of office machinery and computers	2.4
{34} Manufacture of motor vehicles, trailers and semi-trailers	2.4
{15} Manufacturing of food products and beverages	2.3
{90} Sewage and refuse disposal, sanitation and similar activities	2.2
{91} Activities of membership organisations not elsewhere classified	2.1
{21} Manufacture of pulp, paper and paper products	2.1
{28} Manufacture of fabricated metal products, except machinery and equipment	1.9
{99} International organisations and bodies	1.6
{85} Health and social work	1.6
{75} Public administration and defence; social security	1.5

Source: HESA, ad hoc data enquiry (2000a).

6.3. *Policy mechanisms for ISRs in research training*

This section examines the policy measures currently in use in the United Kingdom that aim to support training through industry-science linkages. Table 25 shows the principal schemes currently in operation, the majority of which are government sponsored. There is, however, a small number of influential schemes operated by firms. This section outlines the training activities before going on to concentrate on the measures adopted to promote mobility of research students and the mobility of existing researchers.

Table 25. **UK ISR and policy mechanisms: research training and mobility**

Form of ISR	Policy mechanism
Research training activities and/or centres for graduates and undergraduates	• CASE • Industrial CASE • CASE Plus • CASE for New Academics • Research Masters (MRes) • Integrated Graduate Development Scheme (IGDS) (EPSRC and BBSRC) • Engineering Doctorates • Postgraduate Training Partnerships (PTP) • Science Enterprise Challenge – to bring entrepreneurial skills to science • STEP (Shell Technology Enterprise Programme – funded with DTI) • Chemical engineering (Chem. Eng) studentships • UK Bio-science Business Plan Competition • Biotechnology Young Entrepreneurs Scheme (YES) • Research Councils' Graduate Schools Programme • Enterprise Fellowships (Royal Society) • Shell Technology Enterprise Programme (STEP)
Research personnel mobility schemes	• Teaching Company Scheme (TCS) • Colleges and Business in Partnership (CBP) • Faraday Partnerships • College business partnerships (CBP) (in Wales only) • International Technology Service • Industrial secondments (by DTI for engineering) • Personal research chairs • Senior research fellows • Royal Society Industry Fellowships • Ministry of Defence Research Fellowships • Short Term Fellowship – British Telecom • Higher Education Business Partnership Projects (mostly with Training and Enterprise Councils and SMEs)

Education and training of graduates through ISRs

The production of human capital in the United Kingdom has been strongly influenced over many years by the need to ensure that graduates receive both the theoretical and the practical knowledge required by industry and the professions.

Table 26 shows the extent to which industrial needs are taken into account in the development of curricula for undergraduate and postgraduate courses. Industrial links to undergraduate teaching take a wide variety of forms. At a general level, advisory committees to faculties with courses of vocational relevance typically contain industrial representatives. Active involvement by industry in courses includes provision of visiting speakers (and occasionally lecture series), validation of courses, membership of examination boards, carrying out of student projects in collaboration with firms, and sponsorship of student prizes. Also widespread is a trend towards seeking to develop students' transferable skills relevant to the industrial environment, including for example computer literacy and working in teams. Training for entrepreneurial skills is supported by the Enterprise in Higher Education initiative.

Table 26. **Postgraduate and undergraduate courses designed jointly with industry, by academic area**

	Engineering	Chemical and physical sciences	Information technology	Health and life sciences	Business, management/ accountancy	Total
Masters courses	27%	9%	9%	20%	36%	100%
Undergraduate courses	39%	3%	8%	8%	42%	100%

Source: Howells *et al.* (1998).

Undergraduate degree courses designed to meet the needs of a specific firm or group of firms were, according to the PREST study in 1998 (Howells *et al.*, 1998), offered by around one-third of the higher education institutions surveyed. Of these courses, almost two-thirds were part-time. The PREST study also found that in terms of the distribution across subject areas, engineering had a much higher representation (39%) than for postgraduate and CASE studentships. Business, management and accountancy was the most important discipline, with 42% of courses designed jointly with an industrial/commercial partner.

The PREST report also found that 92% of HEIs responding to the survey had some work and project placement schemes with industry. Nearly three-quarters (73%) of HEIs in the PREST survey had established mechanisms for sandwich courses. Sandwich courses, in which the student typically spends a year working for a company during the course, are very important for some institutions. The University of Ulster, for example, has some form of sandwich arrangement for 50% of its full-time first-degree students, usually for one year.

The PREST study, which closely examined the role and operation of sandwich courses, found that they were becoming increasingly popular, with many more organisations seeking placements for their students in industry. Schools were also moving into the area, increasing competition for places. Respondents to the PREST survey claimed that industry had become increasingly unwilling to provide placements, although the reasons why this should be so are not clear. Respondents claimed that arranging student placements in certain parts of the country was problematic, with London being identified as an area of particular difficulty. There is a proposal to establish a database of companies and students in the capital, supported by eight HEIs, in order to provide a matching agency for placements.

Sandwich courses were seen as giving students a good lead into a job, offering a means of forging links with industry for staff and providing a selling point for courses. Indeed, one institution had suffered because employers were so impressed by its students that they had tempted them to stay and not complete the course. The rising numbers of part-time students creates a new need that may be addressed by the creation of the university for industry.

Shell, in conjunction with DTI, has established a technology enterprise programme in which 1 500 students have been placed in firms around the United Kingdom. During this eight-week placement – what is effectively a mini-sandwich course, each student has tried to devise a solution to a pressing problem faced by the host business.

In relation to sponsorship, 43% of HEIs indicated that they had an established mechanism for sponsorship of undergraduate courses. Although this is lower than for postgraduate courses, it is still a high figure and is likely to continue to grow. In terms of the sponsorship profile regarding the design and implementation of courses, the pattern is very similar to that for masters courses outlined in

Section 5 above, although industry involvement in the latter is somewhat higher. Direct sponsorship of students was generally considered to be quite rare, although there is renewed interest in this activity as a result of the introduction of tuition fees. It was felt that employers might be willing to pay these fees, particularly for final-year students in relevant courses.

Within the existing undergraduate provision, greater emphasis has been placed upon transferable skills and the identification of the types of skill requirements demanded by employers. The government's Life Long Learning Green Paper *The Learning Age* (Department for Education and Employment, 1998) set out a range of proposals on the further development of educational provision to meet the needs of industry. Particular attention was given to learning in the workplace, support for smaller firms which do not have a good reputation for training their workforces, and the establishment of sound and agreed targets for skills and knowledge by setting up a National Skills Taskforce with a new Skills Unit in the Department for Education and Employment.

Foundation Degrees

In the last two years, the government has taken steps to better align educational provision in the United Kingdom with the needs of industry through the creation of a new degree format, the Foundation Degree. The prototype Foundation degrees reflect the growth of the new service economy with 16 degrees in the new media and technologies, and 16 in the areas of health, supporting public services and tourism. The number of engineering and chemical industry related courses is low, with only three indicated in the initial group of around 40 prototype courses. Foundation degrees are not yet significant for industry-science relations as the number of courses operating currently is small. However, if the number of degrees increases, and industry sees advantages from them, their relative importance within the educational sector could change.

Collaborative doctoral awards – CASE

At graduate level, collaborative awards operated through the CASE scheme (Collaborative Awards in Science & Engineering) have increased in number and new forms of collaborative awards have been explored, including CASE for new academics. In this scheme, the student receives a grant from a Research Council in addition to a contribution from an industrial partner. The doctoral research addresses an industrial or commercial problem. Supervision is provided jointly by academic and company representatives.

Other schemes operating to educate graduates through links with industry include engineering doctorates, postgraduate training partnerships, research masters, and the Total Technology Scheme, which is closely linked to the CASE system. The array of initiatives also includes the integrated graduate development scheme, the creation of Research Council Fellowships, and the running of graduate schools for graduates in the second or third year of their doctoral degrees who wish to know more about industrial and commercial careers.

Following the DTI's 1998 Competitiveness White Paper, two rounds of bids under a new Science Enterprise Challenge have been announced. Total funding for schemes which will establish 12 centres of excellence to teach the state of the art in entrepreneurial and business skills to graduates and undergraduates is thought likely to reach GBP 57 million. The 12 centres involve a further 22 academic institutions across the United Kingdom.

At the level of postgraduate education, a range of initiatives are being undertaken autonomously by those responsible for particular subject areas such engineering doctorates, masters of research degree, and the integrated graduate development scheme and through the funding agencies by the Research and Funding Councils. The collaborative mode of research training is also being expanded to provide industrial CASE studentships where industrial partners choose their academic partner organisation.

Teaching Company Scheme (TCS)

The Teaching Company Scheme (now known as TCS), which dates from 1975, is operated by the Department of Trade and Industry. TCS has been regarded as a cornerstone of the development of industry-science relations. The scheme operates through partnerships between firms and academic institutions. Partnerships employ a graduate student (termed an associate) originally with a science or engineering background (although schemes now include social science graduates), who spends 90% of his/her time working in a company on specific projects. The balance of his/her time is spent in the higher education institution where he/she undergoes training.

The purpose of the TCS is "to strengthen the competitiveness and wealth creation of the United Kingdom by the stimulation of innovation in industry through collaborative partnerships between the science, engineering and technology base and industry" (speech by Alan Johnson, Minister of State at the Department of Trade and Industry, 4 December 2000). TCS has a number of precise objectives that ensure close working and collaboration between industry and academia. The *Annual Report* of the Scheme states that its main objectives are to:

- Facilitate the transfer of technology and the spread of technical and management skills, and encourage industrial investment in training, research and development.

- Provide industry-based training, supervised jointly by personnel in the science, engineering and technology base and in business, for high-calibre graduates intending to pursue careers in industry.

- Enhance the levels of research and training in the science, engineering and technology base that is relevant to business by stimulating collaborative research and development projects and forging lasting partnerships between the science, engineering and technology base and business.

Programmes have ranged in size from a single associate employed over a two-year period to a group of 14 associates employed over a period of three years on a contract which has been renewed. Around 2 000 partnerships have been created since the scheme's foundation in 1975. Until 1981, funding for the schemes was entirely out of public funds. Thereafter, the companies themselves have provided up to 60% of the cost of the programmes and at least 50% of the cost of renewed projects. SMEs pay less than larger firms towards the cost of the programmes, usually 30% rather than the 60% which larger firms are required to contribute.

There has been a growing involvement of TCS with smaller companies and in 2000 nearly all the schemes in operation (91%) were with SMEs. Plans are currently being made to increase the number of schemes through a doubling of the budget allocated to the TCS. The government intends to increase the number of active partnerships from around 703 (in April 2000) to 1 000 by the end of 2001. The government is on record as stating[3] that TCS is its premier technology transfer scheme.

The TCS now sees the application of knowledge from a broader range of disciplines than envisaged at the scheme's inception. In 2000, 49% of graduates come from an engineering discipline, 24% from an information technology background, 19% are science based, and a further 9% involve the application of social science knowledge.

The *Annual Report* of the TCS (1999-2000) claims that the scheme has been successful, with the following major achievements attributable, on average, to each partnership:

- A one-off increase in profit before tax of GBP 98 000 (with the highest reported increase being GBP 500 000).

- An annual increase in profit before tax of GBP 138 000 (with the highest reported estimate being GBP 3.5 million).

- Investment in plant and machinery of GBP 157 000.

- 3.5 new jobs created.

- 18.5 company staff trained.

Further co-operation between the higher education institutions and the industrial partners is also a likely outcome of the partnerships, with 78% of firms remaining in contact with their partner institution. Associates tend to remain within the companies once the formal partnership comes to an end. The current cost of government grants to the scheme was around GBP 23 million in the financial year 1999-2000.

6.4. *Research personnel mobility*

The movement of personnel between organisations leads to the transmission of tacit knowledge and has been shown by a large number of studies to be a valuable part of the innovation process. *Talent not Technology* (SPRU, 2000) is the most recent of such studies. For a detailed examination of the nature of the role of interpersonal linkages and geographical clustering. Moreover, despite their immense theoretical refinement and dimensions, many scientific, engineering, medical and technology subjects are deeply, if not wholly, rooted in practical problems and concerns, separation from which exhausts these academic disciplines of their relevance and vitality. Consequently, within many fields of science, links between students, postgraduates and staff, on the one hand, and staff from industry, on the other, have been developed over many years.

Such links often provide the foundations for the academic subjects themselves, in terms of subject definition and governance.[4] Recently, policy makers at all levels, whether within fields of science or at the level of national and international governments, have spurred further interactions through a welter of initiatives. In this area, as in other areas of industry-science linkages, a major issue for evaluation is to distinguish between the effects of schemes that are autonomous – perhaps derived by those responsible for a particular field of science – and those activities which stem from government and which are specifically directed at increasing the level of innovation.

In the area of research mobility of trained academic and industrial staff, there are no legal or contractual barriers that either prevent or facilitate movement. However, barriers to movement between the academic and industrial sectors do exist through salary differentials and pension arrangements. These barriers were sited by respondents to the survey as placing major restrictions on staff movement between academic institutions and industry.

The most important mechanisms for movement are through personal contacts which often give rise to research collaborations. Among the respondents to the questionnaire, the provision of a sabbatical year was thought to be useful means of promoting the exchange of staff; but the Research Assessment Exercise was considered to be a significant barrier to the greater movement of senior university staff to industry because of a lack of an academic publication track record.

In the area of engineering science, professional bodies have played a significant role in ISRs. The Engineering Council and the Royal Academy of Engineering are two of the most active professional bodies in industry-science relationships. The Engineering Council plays a major role in the certification of professional staff and promotion of engineering expertise. Its members and affiliates include a number of the foremost professional bodies in the United Kingdom. The Royal Academy of Engineering works collaboratively with the Council, but also undertakes a number of its own activities. These currently include the following four main themes: to ensure that the supply of trained graduates meets the needs of industry; to promote exchange between industry and science in the process of curriculum development; to ensure that academic researchers are aware of developments faced by the profession itself; and to convey the findings of the latest engineering research directly to the profession at the workplace. The third objective is met through a number of personnel mobility schemes set up by the RAEng or in which it its active as an international participant. The fourth objective is met through a relatively new initiative, the Partnership for Profitable Product Improvement, which is jointly funded with the Department of Trade and Industry.

In respect of personnel mobility, it is the Royal Academy of Engineering's Visiting Professor Scheme that ensures that the needs of industry are understood by engineering students and researchers. This scheme allows senior engineers from industry to spend time in a university. The Royal Academy of Engineering has also been active in seeking greater recognition of the value of practical contributions of academics, publishing a report which proposes significant changes to the current HEFCE research assessment exercise (Royal Academy of Engineering, 2000).

The collaborative award system (CASE), which was established for postgraduate education, has now been introduced for new academic staff as a means of broadening their experience. CASE for New Academics provides an opportunity for academic staff to work within a firm once the period of studentship comes to an end.

7.   Conclusions

7.1.   *Industry-science relations in the United Kingdom*

In common with other OECD countries, the United Kingdom is committed to the development of a knowledge-driven economy where knowledge and know-how form the key foundations of economic success and performance in the global industrial economy. Within the knowledge-based economies, the contribution of the science base is vital in the supply of both knowledge and trained people. A third role has been emerging in which the science and engineering base acts an initiator of entrepreneurial ventures which, after a suitable period of incubation, pass into full industrial ownership in return for financial gain to the institution. However, this last aspect is still rather small compared with the other two.

Taking the three main dimensions of linkage, this review has confirmed the main findings of the PREST survey of 1998:

- *Research grant and contract income* from industry is becoming absolutely and relatively more important over time. Increasingly, industrial income is used in concert with public resources across a range of schemes. However, major success in obtaining such income is concentrated upon relatively few institutions. It is desirable that variety and specialisation should exist among institutions, but the implication remains that there is substantial growth potential if average practice moves closer to best practice.

- *Commercialisation* of the results of academic research hinges upon securing intellectual property rights. There is growth according to all indicators, in terms of patent, licence and option activity, and income. However, there is also concern about the cost of exploiting and protecting IPR.

- *In the context of teaching and training,* relationships with industry are increasing at all levels. It should be stressed that industry regards the supply of trained people as its first priority, even from research collaboration. Postgraduate activity is dominated by policy-led initiatives, notably the TCS and CASE. However, industry is also becoming more involved in the design and implementation of lower-level courses.

7.2.   *Role of framework conditions*

In the United Kingdom, framework conditions are increasingly favourable to the collaboration of industry and science. A change of culture is occurring in response to shifting incentives and there is a growing alignment between framework conditions and industry-science interaction.

However, within the universities, the importance of publication within the Research Assessment Exercise carries a risk that academic work embodying public good characteristics but which is not publishable will continue to be undervalued. A gap exists between the incentives to institutions which are the consequences of national policies, and the communication of those incentives in a way which is meaningful to individual researchers. The problems are increasingly cultural rather than contractual.

Interviewees pointed out that the present range and mix of policies was too extensive and therefore too complicated and that a rationalisation might allow better targeting of initiatives. There is an urgent need for rigorous evaluation of the new initiatives so that policy design can be tuned to the needs of new and growing firms.

7.3.   *Policy issues for the United Kingdom*

To conclude this report, some longer-term issues are raised.

- Commercialisation by spin-offs and licensing of technology has received central attention in research and innovation policy. Such activity is, of course, desirable but the balance of emphasis has distracted attention from the much larger challenge of fostering relations with existing firms, particularly those in more traditional sectors and of a smaller size. These latter categories also need to participate in the knowledge economy if they are to remain competitive – although they may not always be aware of this.

- A relatively large infrastructure of intermediary organisations has developed in response to successive initiatives. These may be part of the main players in ISRs or may exist independently, with a mandate for regional development being a common mission. The issue at stake is whether excessive emphasis on specialised transfer agencies could monopolise knowledge flows and act as a barrier to the creation of a positive knowledge culture diffused throughout the industry-science nexus. In other words, is there a risk in consigning ISRs to peripheral units away from the core?

- Specific policies for the promotion of ISRs may not be sufficient to counter opposing forces. The Research Assessment Exercise, at least in its past formats, has frequently been identified as a barrier to ISR, even if it fulfils a valuable function in terms of its main objectives. On the industry side, the low level of R&D investment in most sectors of UK industry greatly reduces the potential for ISRs. Globalisation has also caused some traditional links to be cut as firms rationalise their R&D activities. On the positive side, the science base is also a major attractor of mobile R&D investment. This is an increasingly important rationale for public support.

- It cannot be assumed that all ISR activities are complementary. The drive to commercialisation by universities or research institutions can easily inhibit other collaboration with industry and lead to conflicts over the ownership of intellectual property. Policies which have caused universities, government laboratories and industry to converge have also meant that previous collaborators become, at least in part, competitors (Georghiou, 1998). Commercialisation and privatisation of the public sector research establishments may challenge existing patterns of industry-science relationships. This is currently a matter of debate as a result of the privatisation of the Defence Evaluation Research Agency, which historically has funded research in universities and defence firms. Industry has raised the question of whether the existing relationships between industry and science in this key area for the United Kingdom may become threatened or disrupted (Molas-Gallart, 2000).

- One final issue for debate concerns the nature of linkages. There is evidence to suggest that these are sector- and field-specific, with links in biotechnology being of a quite different nature to links in the service sector, for example. In the light of this, extrapolation of policies from one sector to another is dangerous.

NOTES

1. Following the general election of 2001, the Department of the Environment Transport and the Regions was restructured to create a new department, the Department of the Environment, Food and Regional Affairs (DEFRA).

2. http://www.innovation.gov.uk/projects/rd_scoreboard/introfr.html

3. http://www.dti.gov.uk/ost/link/tcs.html

4. The Chemistry Department at Imperial College, London, has recently been restructured. The new arrangements bring the internal organisation of the department into closer alignment with the way in which the chemical industry itself carries out research (see PREST, 2000). The new structure comprises the following groups: Analytical Biological and Biophysical Chemistry Catalysis, Advanced Materials, Computational and Structural Electronic Materials, and Interfacial Science Synthesis.

REFERENCES

Baker, J. (1999), *Creating Knowledge, Generating Wealth – Realising the Economic Potential of Public Sector Research Establishments*, HM Treasury, HMSO, London.

Barker, K. and P. Street (1998), "Technology Transfer and Innovation in the Construction Industry: Implications for Public Policy", in C.E. Garcia and L. Sanz-Menendez (eds.), *Management and Technology* 5, COST A3, European Commission, Brussels.

Boden, R., D. Cox, L. Georghiou and K. Barker (2001), "Administrative Reform of United Kingdom Government Research Establishments: Case Studies of New Organisational Forms", in D. Cox, P. Gummett and K. Barker (eds.), *Laboratories: Transition and Transformation*, IOS Press, Amsterdam.

British Venture Capital Association (2000), "Be Bolder", Budget Submission, BVCA, London.

Carter, C.F. and B.R. Williams (1957), *Industry and Technical Progress*, Oxford University Press.

Committee of Vice Chancellors and Principals (2000), *Spin-off and Start-ups in UK Universities: A Report by Sir Douglas Hague and Kate Oakley*, CVCP, London.

Cox, D., L. Georghiou and A. Salazar (2000), "Links to the Science Base of the Information Technology and Biotechnology Industries", report produced on behalf of the Economic and Social Research Council for the Director General of Research Councils, PREST, University of Manchester.

Cunningham, P. and S. Hinder (1999), *A Guide to the Organisation of Science and Technology in Britain*, The British Council, Science Section, Manchester.

Defence Evaluation and Research Agency (2000), *Annual Report 1999/2000*, The Stationery Office, London.

Department for Education and Employment (1998), *The Learning Age – A Renaissance for a New Britain*, http://www.lifelonglearning.co.uk/greenpaper

Department of Trade and Industry (1998), *Our Competitive Future – Building the Knowledge Driven Economy*, Competitiveness White Paper, Cmnd. 4176, HMSO, London.

Department of Trade and Industry (1999a), *Forward Look of Government-funded Science, Engineering and Technology*, Cmnd. 4363, HMSO, London.

Department of Trade and Industry (1999b), *Science, Engineering and Technology Statistics 1999*, Cmnd. 4408, HMSO, London.

Department of Trade and Industry (2000a), *Excellence and Opportunity: A Science and Innovation Policy for the 21st Century*, Cmnd. 4814, HMSO, London.

Department of Trade and Industry (2000b), *Science, Engineering and Technology Statistics 2000*, Cmnd. 4902, HMSO, London.

Department of Trade and Industry (2000c), *The Exploitation and Application of Science*, HMSO, London.

European Commission (2000), *European Trend Chart on Innovation*, European Union Project, DG Enterprise, http://www.cordis.lu/trendchart/.

Faulkner, W. and J. Senker (1995), *Knowledge Frontiers – Public Sector Research and Industrial Innovation in Biotechnology, Engineering Ceramics, and Parallel Computing*, Clarendon Press, Oxford.

Georghiou, L. (1998), "Science, Technology and Innovation Policy for the 21st Century", *Science and Public Policy*, April, pp. 135-137.

Georghiou, L. (2001), "The United Kingdom National System of Research, Technology and Innovation", in P. Larédo and P. Mustar (eds.), *Research and Innovation Policies: An International Comparative Analysis*, Edward Elgar, Northampton.

Gibbons, M. and R. Johnston (1974), "The Roles of Science in Technological Innovation", *Research Policy* 3, pp. 220-242.

Gibbons, M., C. Limoges, H. Nowotny, S. Schwartzman, P. Scott and M. Trow (1994), *The New Production of Knowledge*, Sage Publications, London.

HESA (2000a), Ad hoc information requests, UK Government Higher Education Statistical Agency.

HESA (2000b), "Higher Education On-Line Information System", Directory of institutional financial data, UK Government Higher Education Statistical Agency.

HM Treasury (1999), *Creating Knowledge Creating Wealth: Realising the Economic Potential of Public Sector Research Establishments*, August, HMSO, London.

HM Treasury and Department of Trade and Industry (1998), *Innovating for the Future: Investing in R&D – A Consultative Document*, March.

Howells, J. (1999), "Research and Technology Outsourcing", *Technology Analysis and Strategic Management*, Vol. 11, No. 1, pp. 17-29.

Howells, J., M. Nedeva and L. Georghiou (1998), "Industry-Academic Links in the UK", Report to the Higher Education Funding Councils for England, Scotland and Wales, PREST, University of Manchester.

Langrish, J., M. Gibbons, W.G. Evans and F.R. Jevons (1972), *Wealth from Knowledge – A Study of Innovation in Industry*, MacMillan, London.

Martin, B. and A. Salter (1996), *The Relationship between Publicly Funded Basic Research and Economic Performance – A SPRU Review*, Science Policy Research Unit, Report prepared for HM Treasury, July.

Massey, D., P. Quintas and D. Wield (1992), "Academic Industry Links and Innovation: Questioning the Science Park Model", *Technovation* 12, pp. 6-75.

Molas-Gallart, J. (2000), "Government Defence Research Establishments: The Uncertain Outcome of Institutional Change", SPRU Electronic Working Paper No. 47.

OECD (2000), *OECD Science, Technology and Industry Outlook 2000*, OECD, Paris.

OST (1993), *Realising Our Potential – A Strategy for Science, Engineering and Technology, White Paper*, HMSO, London.

OST (2000), *Good Practice for Public Sector Research Establishments on Staff Incentives and the Management of Conflicts of Interest*, July, Office of Science and Technology, UK Department of Trade and Industry, http://www.dti.gov.uk/ost/aboutost/psre.htm

PREST (2000), "Report on the Research Assessment Exercise for HEFCE", PREST, University of Manchester.

Royal Academy of Engineering (2000), *Measuring Excellence in Engineering Research*, http://www.raeng.org.uk/policy/reports

Scottish Higher Education Funding Council (2001), *Annual Report*, http://www.shefc.ac.uk

SPRU (2000), *Talent, Not Technology: Publicly Funded Research and Innovation in the UK*, http://www.sussex.ac.uk/spru/news/talent.html

Stewart, G. (1999), *The Partnership Between Science and Industry*, The British Library, London.

United Kingdom Science Park Association (1998), *Annual Report*, UKSPA, Birmingham.

Chapter 6

INDUSTRY-SCIENCE RELATIONSHIPS IN JAPAN*

1.   Introduction

Japan has now recognised the growing importance of science-based knowledge in industrial R&D and innovation. Industry-science relationships (ISR) are currently undergoing deep-seated reform, with the ensuing amendment of the rules that had previously posed restrictions on the activities of national universities and research institutes. Furthermore, a new legal status has been created to cover the national research institutes (NRIs) and other government-affiliated bodies, and the majority of these institutions will soon become "independent administrative legal entities" (*dokuritsu-gyousei-houjin*). In addition, discussions have recently begun on the question of transforming the national universities into independent administrative legal entities. Based on the available data, this chapter presents the current situation of ISR in Japan.

1.1.   The increasing importance of science and knowledge in industry

As indicated in Part 1, the following characteristics have been observed by Narin *et al.* (1997) on the basis of an analysis of patent registration data in the United States:

- Citations of scientific papers in US patents have increased.

- Recent research papers tend to be cited more frequently.

- Patent applicants show a tendency to cite research papers from their own countries.

- Almost three-quarters of the patents registered by US firms cited scientific papers relating to research carried out by universities and NRIs using government funds.

Accordingly, the relationship between patents and the results of basic research as represented by scientific papers is becoming stronger. The key role played by public institutions such as universities and NRIs in the creation of basic research results is becoming increasingly apparent.

Science and Technology (S&T) Indicators, published by the Japanese National Institute of Science and Technology Policy (NISTEP, 2000), analyses US patent registration data according to applicants' nationalities. The report finds that, although the science-linkage of patents filed by

*   This chapter has been written by Ryuji Shimoda and Akira Goto, Institute of Innovation Research, Hitotsubashi University, Japan. The authors wish to express their appreciation to researchers and staff of universities and national research institutes and government officials who were interviewed and/or provided information.

Japanese nationals in US patents is relatively weak, it has become stronger over the years. Taking areas in which there are strong linkages, a similar tendency can be observed in patents registered by Americans and in those registered by Japanese nationals. The science-linkage is comparatively strong in the life sciences for patent applications by both Japanese and American nationals (Figure 1). This trend points to the importance of science and of the national institutions that conduct scientific research in a country's innovation system.

Figure 1. **Science linkages in manufacturing**

a) Japan: selected fields

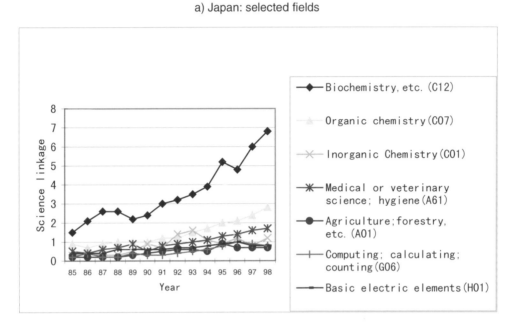

b) United States: selected fields

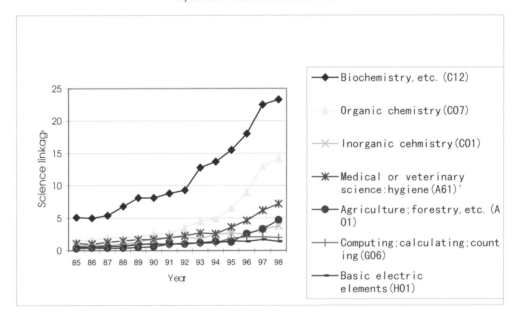

Source: NISTEP (2000), *S&T Indicators - 2000,* based on CHI Research Inc., *National Technology Indicators Database.*

1.2. R&D activities in Japan

This section provides an overview of Japanese R&D activities before discussing the state of ISR in Japan.

The ratio of total R&D expenditure to gross domestic product (GDP) in Japan exceeds 3%. This represents the highest ratio among the world's major industrialised countries. In 1998, total Japanese R&D expenditure on the natural sciences was JPY 14 824 471 million (JPY 16 139 925 million, if the social sciences and humanities are included). In terms of R&D expenditure, Japan ranks second only to the United States.

Regarding sources of R&D funding in Japan, 78.0% comes from the private sector, 21.7% from government and 0.3% from foreign sources. The significant contribution of the private sector to research funds is immediately apparent from these figures. The government share fell to less than 20% for several years in the late 1980s and early 1990s owing to active private sector R&D funding and strict fiscal policy. It recovered somewhat during the late 1990s as a result of increases in government spending coupled with sluggish private sector expenditure (in addition to industry, private universities are included in the above description of the private sector).

The costs of R&D activities performed by firms are borne almost entirely by the firms themselves. The flow of funds from government to industry is small. A mere 2% of total research spending by industry (JPY 223 400 million) is financed by public funds. In the natural sciences, research funds flowing from industry to academia represent JPY 59 400 million, corresponding to 3% of the total research funds (JPY 2 012 200 million) spent by universities on R&D (this figure includes private universities). For reasons that will be explained later, the flow of funds from industry to government research institutes is very small. An analysis of the flow of funds between the various institutions alone indicates that ISR in Japan is not yet well established.

In April 1999, Japanese researchers numbered 733 000. However, on a full-time equivalent (FTE) basis, NISTEP estimates put the number of researchers at 480 000. The distribution of researchers by type of institutions is as follows: industry (58.6%); government research institutes (4.2%); universities (35%); and private research institutes (2.2%). For comparative purposes, the number of researchers per 10 000 population of the major OECD countries is as follows: 57.9 (FTE estimate: 38) in Japan (1999); 37.5 in the United States (1995); 28.3 in Germany (1995); 26.5 in France (1996) and 24.8 in the United Kingdom (1996).

Funding for research is broken down into that for basic research, applied research, and development, according to its purpose. The shares in Japan in 1998 were 13.9%, 24.6% and 61.4%, respectively. Total basic research funding amounted to JPY 2 139 520 million, of which JPY 600 746 million was spent by private industry, JPY 332 312 million by government research institutes, JPY 95 603 million by private research institutes and JPY 1 110 859 million by universities. It is noteworthy that in Japan industry benefits from a relatively large percentage of the nation's total basic research expenditure, compared with other major R&D-performing countries.

1.3. Publication of research papers

Publication of scientific papers is one of the major indicators of R&D output. The number of research papers published by Japanese researchers has grown steadily in recent years, in terms of both volume and share of the world's publications. Data on the publication of research papers (social sciences excluded) in the five-year period ending 1998 are compared with the five-year period ending

161

1988 using SCI data in Figure 2. The growth rate of the whole SCI database (social sciences excluded) between the two periods was 36.7%. The corresponding growth rates were: 72.7% for Japan, 60.6% for France, 50.9% for Germany, 42% for the United Kingdom and 28.9% for the United States. Japan had the highest growth rate among the five major industrialised countries. Moreover, in terms of the share of the number of scientific publications, Japan has ranked second, following the United States, since 1990. In 1998, Japan published 66 000 papers (10.3%) compared to 211 000 by the United States (32.8%) of a total of 643 000 (NISTEP, 2000).

Figure 2. **Publications by major industrialised countries**

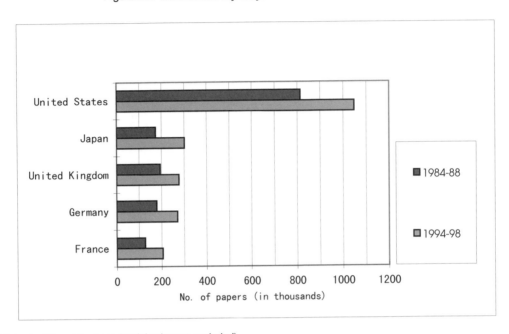

Note: Total scientific publications (social sciences excluded).
Source: NISTEP (2000), *S&T Indicators - 2000.*

1.4.   Researchers in universities and government research institutes by field of science

One indicator used to provide a clearer picture of the areas of research conducted by universities and government research institutes is the distribution of researchers among organisations and fields of science. It should be borne in mind that the numbers of researchers are not presented on a full-time equivalent basis but as simple head-counts.

Table 1. **Distribution of university researchers by type of organisation and field of science**

At 1 April 1999

Field of science / Organisation		Humanities and social sciences	Natural sciences					Others	Total
			Sub-total	Physical science	Engineering	Agriculture	Health		
National	Number of researchers	14 323	93 037	13 580	30 240	8 289	40 928	13 099	120 459
	Share of total	6%	36%	5%	12%	3%	16%	5%	47%
Public (local gov't)	Number of researchers	3 448	12 837	1 387	2 218	712	8 520	1 306	17 591
	Share of total	1%	5%	1%	1%	<1%	3%	1%	7%
Private	Number of researchers	41 209	60 984	1 828	14 673	1 725	42 758	16 197	118 390
	Share of total	16%	24%	1%	6%	1%	17%	6%	46%
Total	Number of researchers	58 980	166 858	16 795	47 131	10 726	92 206	30 602	256 440
	Share of total	23%	65%	7%	18%	4%	36%	12%	100%

Source: Management and Co-ordination Agency (MCA), *Report on the Survey of Research and Development.*

Universities

The numbers of researchers, classified by organisation/faculty are shown in Table 1. Although it would appear that large numbers of university and college researchers are engaged in health R&D, this is because, in Japan, doctoral students and members of the medical offices (a sort of trainee) are included in the definition of researchers. National universities occupy a greater share as far as engineering and the physical sciences are concerned.

Central government research institutes

Central government research institutes consist of national research institutes and special public research corporations. The distribution of central government research institute researchers by type of organisation and by field of science is shown in Table 2.

Table 2. **Distribution of researchers at central government research institutes by type of organisation and field of science**

At 1 April 1999

Field of science / Organisation		Humanities and social sciences	Natural sciences					Others	Total
			Sub-total	Physical science	Engineering	Agriculture	Health		
National research institute	Number of researchers	490	10 784	1 616	4 573	3 145	1 450	197	11 471
	Share of total	3%	66%	10%	28%	19%	9%	1%	70%
Special public research corporation	Number of researchers	250	4 613	3 843	708	62	0	0	4 863
	Share of total	2%	28%	24%	4%	<1%	0%	0	30%
Total	Number of researchers	740	15 397	5 459	5 281	3 207	1 450	197	16 334
	Share of total	5%	94%	33%	32%	20%	9%	1%	100%

Source: MCA, *Report on the Survey of Research and Development.*

Researchers in central government research institutes number 16 000, a figure much lower than that for researchers at universities. In terms of organisation type, 70% of researchers are affiliated with national research institutes, while 30% are employed in special public research corporations. The natural sciences employ the majority of researchers. Engineering-based research is the major activity of the national research institutes, while about 20% is targeted at agriculture. Many researchers are employed in special corporations specialising in the physical sciences, mainly because of the significant number of researchers affiliated with the large-scale Japanese Atomic Energy Research Institute (JAERI) and the Institute of Physical and Chemical Research (RIKEN).

Local government operated research institutes

Government research institutes include those operated by local governments. The number of researchers in local government operated research institutes is 15 000, and more than half of them are carrying out research related to agriculture (Table 3).

Table 3. **Distribution of researchers at local government operated research institutes among organisations and fields of science**

At 1April 1999

| Organisation | Field of science | Humanities and social sciences | Natural sciences | | | | | Others | Total |
			Sub-total	Physical science	Engineering	Agriculture	Health		
Local government	Number of researchers	39	13 657	1 173	3 369	7 646	1 469	880	14 576
	Share of total	<1%	94%	8%	23%	52%	10%	6%	100%

Source: MCA, *Report on the Survey of Research and Development.*

1.5. *Government support for higher education*

Total public educational expenses amount to JPY 23 895 800 million, or 4.31% of 1996 GNP. The higher education sector, at JPY 3 575 431 million, accounts for 15% of total public expenditure on education (0.65% of GNP) (Ministry of Education, 1999a). In recent years, the share of higher education in total public support to education has increased steadily, while the ratio of public support to higher education to GNP has remained fairly stable.

Table 4. **Public support to higher education in Japan**

	Total public educational expenses (JPY million)	Public expenses for higher education (JPY million)	Percentage of higher education (%)	Ratio of total public educational expenses to GNP (%)	Ratio of public expenses for higher education to GNP (%)
1992	22 334 916	2 989 108	13.4	4.69	0.63
1993	23 050 901	3 343 723	14.5	4.79	0.69
1994	23 043 851	3 381 491	14.7	4.38	0.64
1995	23 766 348	3 614 650	15.2	4.31	0.66
1996	23 895 790	3 575 431	15.0	4.31	0.65

Source: Ministry of Education (1999a), *Education Policy in Japan – 1999.*

2. Industry-science relations in Japan: university, government research institute and industry co-operation

In the following sections, universities (mainly national universities) and the national research institutes will be discussed (henceforth, the term "NRI" refers to research institutes directly attached to ministries or agencies of the national government).

Generally speaking, special public research corporations and local government operated research institutes are included among government research institutes. However, special public research corporations generally conduct R&D in fields limited to such areas as atomic energy or space. Their relationship with industry is, thus, not typical of the usual relationship between industry and government research institutes. As such, the special public research corporations are largely excluded from the description of ISR presented in this chapter. Furthermore, since more than half of the researchers at local government operated research institutes undertake agriculture-related research (for which data are generally scarce), these institutions will also be excluded from further description of ISR in Japan.

2.1. Flow of human resources (education and training)

Supply of graduates and postgraduates to industry

One of universities' key contributions to industry lies in the supply of human resources in terms of graduates and postgraduates. Table 5 illustrates the evolution of employment among graduates and postgraduates.

Table 5. **Employment of graduates and postgraduates, 1995-99**

Number of graduates and postgraduates

	March 1995	March 1996	March 1997	March 1998	March 1999
Bachelor course graduates	330 998	337 805	349 241	347 549	320 072
Masters course graduates	28 019	31 747	34 124	35 612	34 190
Doctoral course graduates	4 984	5 621	6 162	6 655	7 113

Source: Ministry of Education (1999b), *School Basic Survey – 1999*; Ministry of Education (1999c), *Statistical Abstract of Education, Science, Sports and Culture – 1999*.

Although the number of new hires fluctuates depending on such factors as the size of the student population and rates of promotion to higher institutions, the number of masters and doctoral graduates finding employment has been on the rise. Table 6 shows the employment situation of graduates by industrial classification in 1999.

2.2. Contract research and donations: transfers of funds

The *Report on the Survey of Research and Development*, published by the Management and Co-ordination Agency (MCA), shows that JPY 59 400 million in research funds were transferred from private firms to universities in the field of the natural sciences in 1998. The flow of funds from industry to university grew until the early part of the 1990s before dropping off in 1993, since when, the amount has remained relatively stable (Table 7).

Table 6. **Employment of all graduates, by industry, March 1999**

Number of graduates

	Total	Humanities	Social science	Physical science	Engineer-ing	Agriculture	Health	Others
Construction								
- Undergraduates	17 252	1 053	3 750	123	10 881	502	2	941
- Masters	1 572	5	26	34	1 369	61		77
- Doctors	91	1	2	10	72	2		4
Manufacturing								
- Undergraduates	62 968	7 243	21 832	2 198	23 491	2 578	1 154	4 472
- Masters	17 685	63	340	1 626	13 580	834	713	529
- Doctors	962	1	23	137	613	68	85	35
Electricity/gas/heat supply/water								
- Undergraduates	1 191	72	664	21	358	14	3	59
- Masters	595		24	10	547	3		11
- Doctors	22				21			1
Transportation/ communication								
- Undergraduates	11 809	3 115	5 974	290	1 389	123	7	911
- Masters	1 272	21	64	138	964	18	4	63
- Doctors	49	2	8	6	30			3
Wholesale/retail, restaurants								
- Undergraduates	70 135	12 949	42 742	812	4 920	1 821	1 541	5 350
- Masters	657	43	129	57	225	74	74	55
- Doctors	19	1	10	1	4	1	2	0
Finance/insurance								
- Undergraduates	31 152	6 661	20 885	404	725	310	18	2 149
- Masters	471	12	225	72	129	11		22
- Doctors	36		26	4	3	1		2
Real estate								
- Undergraduates	4 455	759	2 683	21	502	37	3	450
- Masters	90		25	2	51	2		10
- Doctors	1			1				0
Services								
- Undergraduates	97 537	15 742	33 646	4 103	17 359	2 383	7 390	16 914
- Masters	9 177	624	1 234	896	2 726	361	692	2 644
- Doctors	5 186	294	410	439	846	230	2 677	290
Public services								
- Undergraduates	15 887	2 109	7 988	359	1 894	966	423	2 148
- Masters	1 966	120	424	152	660	280	90	240
- Doctors	436	15	23	74	162	90	48	24
Other								
- Undergraduates	7 733	885	3 428	355	1 459	664	30	912
- Masters	811	57	141	77	281	100	18	137
- Doctors	318	15	22	76	92	34	61	18
Total								
- Undergraduates	320 119	50 588	143 592	8 686	62 978	9 398	10 571	34 306
- Masters	34 296	945	2 632	3 064	20 532	1 744	1 591	3 788
- Doctors	7 120	329	524	748	1 843	426	2 873	377

Source: Ministry of Education (1999b), *School Basic Survey – 1999.*

Table 7. **Trends in R&D funding from industry to universities**

JPY million

	Total funds	National universities	Public universities and colleges	Private universities
1986	25 974	19 231	687	6 056
1987	29 584	22 450	816	6 317
1988	36 725	26 824	901	9 001
1989	39 391	28 879	917	8 595
1990	45 244	33 375	1 368	10 503
1991	49 152	35 701	1 668	11 782
1992	55 845	40 169	2 276	13 399
1993	56 389	40 187	2 473	13 729
1994	53 098	37 279	2 736	13 083
1995	57 698	40 112	2 336	15 250
1996	56 408	39 293	2 729	14 387
1997	60 384	42 584	2 734	15 067
1998	59 375	40 436	2 719	16 221

Source: NISTEP (2000), based on MCA, *Report on the Survey of Research and Development.*

Owing to the recession which occurred in 1992, research funding by the Japanese private sector decreased for the first time since 1959, the launch year of the R&D statistics survey. After 1992, it continued to fall for two consecutive years, before starting to increase in 1995. However, the flow of funds from industry to university has remained fairly stable. This may reflect industry's growing frustration with the Japanese university system. In contrast, the flow of research funds from Japanese firms to foreign institutions have been growing since 1995 (see Section 2.7 below).

Transfers of funds from firms to universities take two forms: contract research and contributions/donations. The following section investigates the situation with the national universities and NRIs, for which data are reasonably robust.

National universities

Private sector contract research

National universities may be entrusted with research contracts by private sector firms, ministries/government agencies or other organisations wishing to optimise their R&D output by drawing on the basic knowledge and talent concentrated in the national universities. Researchers at national universities perform the research specified in the contract, with the expenses incurred being borne by the contracting firm or organisation. The research results are reported to the body providing the funding. Until recently, the results of such R&D, including patents, were owned by the national universities. Recently, however, the rules have been modified to allow part (up to half) of the rights to patents originating from the contracted research to be transferred from the university to the contracting firm or organisation. In addition, the funding firms/organisations retain preferential rights for a ten-year period following the patent application.

Figure 3. **Contract research at national universities**

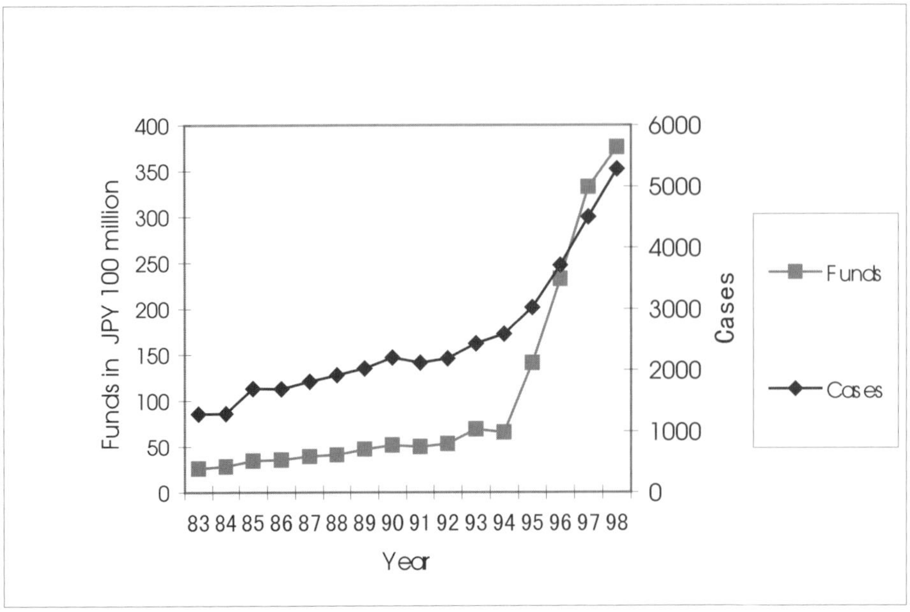

Source: Ministry of Education, *Education Policy in Japan - 1999*; 1998 data from *Japanese Scientific Monthly*, Vol. 5, No. 1.

Recent trends in contract research are illustrated in Figure 3. Since 1995 the total amount of contract research funds has grown rapidly as national universities have begun accepting research grants from government-affiliated organisations such as the JSPS (Japan Society for the Promotion of Science, *Gakujutu-shinkoukai*), the JST (Japan Science and Technology Corporation, *Kagakugijutu-shinkou-jigyoudan*), the NEDO (New Energy and Industrial Technology Development Organisation), and similar bodies. Therefore, the figure does not necessarily imply a rapid increase in funds from industry. For 1995, it was reported that some JPY 7 000 million of funding (or 152 cases), originated from the new research grant systems (Ministry of Education, 1997).

Companies contracting out research to universities are expected to pay the indirect expenses incurred by the universities (*e.g.* technology fees, depreciation of equipment), in addition to the direct research expenses. When a university enters a research contract with a firm or government-affiliated organisation, 30% of the direct expenses (10% in the case of the new research grant systems) are counted as indirect expenses. Since in the past all indirect expenses were paid to the Treasury and not to the university conducting the research, research contracts were not very popular among universities and researchers. As of 2000, however, part of the indirect expenses, together with the licensing revenue from patents, may be distributed to the universities as "research co-operation promotion and incentive expenses". Accordingly, JPY 1 997 million was appropriated for the fiscal 2000 budget.

Since the evaluation of university professors and lecturers is mainly based on their research achievements and on the publication of scientific papers, and does not take into account contributions to industry, the system for contract research prior to 2000 provided little incentive for academics to forge links with industry. Under the new scheme, research funds will increase as the number of research contracts with industry rises. This increase in funds may allow researchers to perform additional curiosity-driven research of their own. Although the amounts involved are small for the time-being, the new scheme should provide university researchers with greater incentives to co-operate with industry.

National universities are government institutions. The National School Special Account controls the national universities' revenue and expenditure. Budget expenditures have to be made within the corresponding fiscal year (April-March of the following year). Until recently, the strict interpretation of the budgetary rules and regulations precluded research contracts exceeding one fiscal year. Contracts for contract or joint research programmes had to be concluded annually, thus placing a heavy administrative burden on firms and university staff. Therefore, contract research was not appropriate for projects requiring speedy implementation. Following criticism of this rigid system, contracts extending for more than one year became possible from FY 2000.

Furthermore, in terms of organisational characteristics, there is a serious imbalance between the national and private universities. Since private universities are in the private sector, contract research constitutes a for-profit endeavour and is thus taxable, while this is not the case for the national universities since they are considered part of the government infrastructure. The point here is that, although the national and private universities perform similar functions, their tax treatment differs.

Shogaku-kifu-kin donations

The national universities accept contributions from firms, individuals and other bodies for research and educational purposes. This is called *Shogaku-kifu-kin* (literally, donation for the promotion of research). The bulk of the funds awarded by firms to national universities come under the *Shogaku-kifu-kin* donation system (recent trends are shown in Figure 4). These funds grew steadily until the beginning of the 1990s, before levelling off. While total research funds provided by the Japanese private sector have been increasing since 1995, *Shogaku-kifu-kin* donations have remained stable and may even be falling. This situation suggests that the private sector may have become frustrated with the universities, as mentioned above.

Figure 4. ***Shogaku-kifu-kin* donations awarded to national universities**

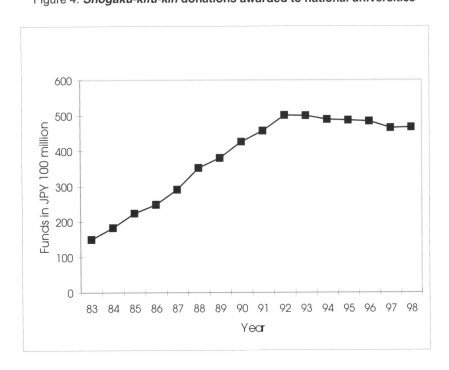

Source: Ministry of Education, *Education Policy in Japan – 1999*; 1998 data from *Japanese Scientific Monthly*, Vol. 5, No. 1.

Donors may attach conditions to their donations concerning such aspects as the subject of the research and the researchers involved. Donors may receive brief reports on the results of research they support. However, conditions relating to the transfer of ownership of intellectual property rights to donors cannot be specified. Donors should not expect direct returns from their donation. Moreover, the results of the research are handled according to the rules and regulations applying to ordinary university research. In general, the invention belongs to the researcher who invented it (see Section 2.4).

The tax system offers preferential treatment; the total contribution is tax deductible for the donor firm. For individuals, the amount of the contribution is deducted from their total income, thus decreasing their taxable income.

Under the Japanese budgetary system, expenditure items (*himoku*) are established defining specific use of the budget item. It is not generally possible to use the budget items for purposes other than specified. In the case of *shogaku-kifu-kin* donations, there is greater flexibility, and researchers are attracted to the system because of the flexible nature of expenses. From the firms' standpoint, maintaining relations with researchers, obtaining access to research results, and the prospect of securing promising graduates seem to be among the main incentives.

Contributions of chairs and research sections

Shogaku-kifu-kin donations can be used to establish chairs for education and research and a section for research in the field designated by the donor. In June 1997, 48 of the endowed chairs and 13 of the endowed research sections in 26 national universities were reported as such (*Education Policy in Japan – 1997*). Although there are some non-profit organisations, most donors are private firms. Endowed chairs and research sections may be said to be one of the most effective means of ensuring that education and/or research in areas of interest to the private sector is carried out in national universities. Under this system, there is no preferential access to research results by contributing firms as the funding is considered as a donation (see above.)

NRIs

Private sector contract research

In 1995, 236 NRIs conducted contract research with private firms and other entities, compared to 501 in 1996, 407 in 1997 and 325 in 1998. More than half were contracts concluded with NRIs attached to the Ministry of Welfare and to the Ministry of Agriculture, Forestry and Fisheries. Total funding amounted to JPY 1 076 million in 1997 and JPY 5 026 million in 1998.[1]

The NRIs carried out less contract research, mainly because of the rigidity of the budget system; in terms of funding, their share represents less than 1% of total research. The research expenses of NRIs should be budgeted in the general government account. However, for the last 20 years or so, the totality of the expenditure budget of the general account of the government has been suppressed (under the so-called "ceiling system"). Generally, funds received from firms for contract research go to the Treasury. However, in the special account, a certain amount of flexible expenditure budgeting at national universities may be possible. Even when the NRIs are able to obtain funding from outside organisations, this does not generally increase their expense budget. To secure the budget for contract research, the NRIs are obliged to reduce their own research budgets. It is not surprising that they have shown little enthusiasm for contract research.

In addition, the NRI administration and researchers tend to favour their own research over contract research requested by other organisations. Greater importance is attached to research carried out by and for their own laboratories. In particular, since a limited number of personnel are employed at the NRIs, there are insufficient human resources for undertaking contract research, and such research is therefore not viewed favourably by the institutes or by the administration. These restrictions are expected to be relaxed once the NRIs become independent administrative legal entities (see Section 2.6).

Donations

While donations provide a source of income for the Treasury, they do not increase the research funds of the NRIs unless a corresponding amount is budgeted in the NRI's expenditure budget. Budgetary increases are made difficult by the ceiling system described above.

2.3. *Joint research by universities, NRIs and industry*

In addition to the supply of graduates to industry, receiving donations from industry and conducting contract research, the joint research conducted by universities, NRIs and industry is an important channel of industry-science relationships. The different types of joint research are described below.

Joint university-private sector research

Joint research conducted with the national universities and industry takes the form of co-operative research projects on subjects of mutual interest. Based on a formal contract, researchers in universities and firms work together on an equal footing. The joint research system was formally launched in 1983. Trends in the number of joint research projects are shown in Figure 5.

Figure 5. **Joint research at national universities**

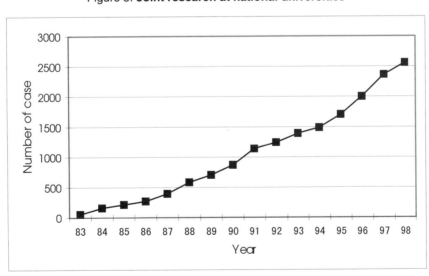

Source: Ministry of Education, *Education Policy in Japan – 1999*; 1998 data from *Japanese Scientific Monthly*, Vol. 5, No. 1.

Joint research takes two forms. In the first case, research is carried out on a university campus by researchers from the university and the firm involved. In the second case, research is conducted both on the university campus and in a firm according to a mutually agreed division of the research.

Although data on the research expenses incurred by the private sector are lacking, according to *Japanese Scientific Monthly* the national university budget for joint research projects amounted to JPY 7 368 million in 2000 (JPY 6 759 million in 1999).

Budget expenditures have to be made within the corresponding government fiscal year. Until recently, joint research contracts (or contract research) exceeding one fiscal year were not possible. Contracts for joint research were heavily burdened by red tape as they had to be rewritten each year, with the exception of short-term contracts which were scheduled to end within the fiscal year in which they started. However, this system was revised to enable the establishment of contracts extending for more than one year.

The above statistics concern only joint research projects with formal contracts. However, it would appear that a large number of co-operative research projects are undertaken which are similar to joint research projects but are not based on formal contracts.

Joint NRI-private sector research

Table 8 describes recent trends in joint research between NRIs and the private sector.

Table 8. **Trends in joint research between NRIs and the private sector**

	1995	1996	1997	1998
Number of contracts	1 497	1 361	1 838	1 994
Number of private sector organisations involved	910	667	1 053	1 141

Source: STA.

The NRIs belonging to the Agency for Industrial Science and Technology (AIST) of the Ministry of Economy, Trade and Industry – METI (formerly MITI) are considered to be comparatively active in co-operating with the private sector. It should be noted, however, that only the number of projects is available and that the amount of research expenditure and number of researchers involved is unknown. In 1998, the AIST research institutes taken as a whole carried out 776 joint research projects. The total number of partnering organisations was 941, of which 453 were private firms.[2]

The partnering organisations fall into three types: half are private firms; about 30% are classified as "other organisations"; and about one-eighth are universities. The AIST NRIs may be divided into two groups: the laboratories at Tsukuba Science City, which play the role of national research centres in their respective fields; and the NRIs clustered in regional blocs. The NRIs in regional blocs tend to co-operate with firms in the region and with local governments. The NRIs in Tsukuba tend to co-operate with large firms and "other organisations" (including JST, NEDO and similar organisations). Research support provided by these organisations takes the form of joint research as a means of avoiding complicated accounting and budgeting procedures.

In addition to the joint research system described above, other research systems aim to promote co-operation between universities, NRIs and industry. The following programmes are examples of the major systems in which university, NRI and industry researchers have forged close links.

ERATO

The ERATO programme is administered by the JST. Under this programme, a research team comprising between ten and 20 researchers is set up under a project leader (PL). The research period is limited to five years, during which time the research team carries out fundamental research and is then dissolved. The scale of the research funding amounts to several hundred million yen per year (including the personnel expenses of the researchers). The PL is selected from among the researchers of the participating universities, NRIs or firms. Until now, many PLs have originated from the universities. The PL selects the researchers who will participate in the project, all of whom are employed by JST. When a researcher employed by a firm participates, a form of secondment from the firm – a temporary transfer – is undertaken. Researchers on secondment from firms accounted for some 40% of all ERATO researchers (Kusunoki, 1997).

ERATO carries out its activities under the auspices of the JST. Research results belong to both its researchers and the JST. However, researchers on secondment from firms who are involved in a project obtain experience and know-how which they are can share with their firms upon their return. Hence, participating firms and academic institutions are effectively able to mutually benefit from the research programme through the knowledge and experience exchanged in the course of the joint work.

Industry Science and Technology R&D Programme

The Industry Science and Technology R&D programme was launched in 1996 with the aim of consolidating the Programme for Fundamental Technology for Next Generation Industry (started in 1981) and the Large-scale Industrial Technology Development Programme (started in 1966), together with a number of smaller programmes. Research contracts are awarded by AIST or NEDO to technology research associations (TRAs) or non-profit organisations for specific R&D projects. A TRA usually brings together researchers from private firms and NRIs engaged in similar research activities.

University researchers may also participate in these projects. However, they rarely conduct research at a TRA on a full-time basis and their role is usually limited to performing research contracted out to universities from a TRA, or to providing technical advice for a TRA.

A new element of this programme, the University-Industry Co-operative Industrial Technology R&D Project, was instituted in 1998. The project aims to promote commercialisation of knowledge produced by universities. The R&D is supported by the Ministry of Education and METI. Joint research on specific themes is carried out on the university campus by university and industry researchers (researchers from private firms first become researchers of non-profit organisations or TRAs and are then sent to universities). In 1999, the budget allocated by the Ministry of Economy, Trade and Industry amounted to JPY 3 568 million.

2.4.   Ownership of intellectual property rights by universities and NRIs, and promotion of patenting

Patent application procedures in universities and NRIs

The following section describes the patent application procedures relating to inventions produced in universities and NRIs.

National universities

In the case of inventions generated by researchers of national universities, the Invention Committee of the respective university will decide whether the intellectual property belongs to the university (*i.e.* the central government) or to the individual researchers. The intellectual property rights to inventions produced by researchers in national universities are generally granted to the individual researchers, except in the case where special research funds and/or a special research facility was used.

Researchers are required to document their inventions, and each application is deliberated by the Invention Committee. General principles for the decision are as follows.

If an invention is generated by: *i)* a research project for application and/or development with special funding provided by the government; or *ii)* a research project for application and/or development in which special facilities are used, regardless of whether or not the special funding is provided by the government, the rights to the invention belong to the government. The government may also acquire the rights to an invention in cases where the researchers request the transfer of their rights to the government.

In all cases except those described above, the intellectual property belongs to the researchers who made the breakthrough. In 1998, national university Invention Committees debated 1 059 cases. In 189 cases, it was judged that the intellectual property belonged to the universities, while in 825 cases, the rights were granted to individual university researchers. In 45 cases, rights were granted by the researchers to the universities (Table 9).

Table 9. **Handling of inventions in national universities, 1998**

Number of cases deliberated by Invention Committees	Intellectual property belongs to government (university)			Intellectual property belongs to inventor
	Granted to the university	Transferred from the inventor	Total	Total
1 059 cases (100%)	189 cases	45 cases	234 cases (22.1%)	825 cases (77.9%)

Source: Ministry of Education.

In cases where the intellectual property rights are retained by the government, the government organisation files the patent applications using public funds. In the past, the JSPS was in charge of patent applications, but this responsibility has recently been taken over by the JST.

Patent applications by national universities are shown in Table 10. The number of applications has increased in recent years as national universities have received grants from new research programmes (see Section 2.1). The inventions generated by research supported by such grants are categorised as "granted to the university".

Table 10. **Patent applications by national universities**

	1994	1995	1996	1997	1998	1999
Number of national patent applications	26	25	35	75	138	191
Number of foreign patent applications	21	23	22	34	73	93
(Number of inventions)	(15)	(15)	(11)	(21)	(32)	(51)

Source: JST.

NRIs

Since inventions by researchers at NRIs are produced in the course of their usual research activities, the intellectual property is deemed to belong to the government. Recently, revisions have been made to the system in the case of the NRIs attached to major ministries and agencies, with the aim of promoting joint ownership by government and researchers. The number of cases in which researchers are attributed a share in the rights has grown (Table 11). Sales of intellectual property and technological assistance to the private sector are expected to lead to greater technology transfer and increased commercialisation. However, firms have to secure government consent before applying the patent/invention since the government retains a share in the intellectual property. Firms may be surprised at the time required for the government to take a decision.

Ownership of patents by individual researchers

The share of patents owned by individual researchers at national universities and NRIs is shown in Table 11.

Table 11. **Patents owned by individual researchers at universities and NRIs**

Percentages

	1995	1996	1997	1998	1999
Share of patents owned by national university researchers	89.7	85.3	83.2	77.9	n.a.
Share of domestic patents owned by NRI researchers	0	2.2	14.9	14.4	(16.8)
Share of overseas patents owned by NRI researchers	0	0	4.7	12.4	(14.2)

1. Share of total patents filed by NRIs partly owned by researchers or full titles to inventions announced by national universities.
2. 1999 = 1 April-31 December.
Source: STA (1999), *Annual Report on Science and Technology Promotion in 1999.*

The responsibility for filing patent applications falls to the individual researchers of the national universities when the intellectual property is assigned to the researcher. University researchers are reluctant to file patent applications because they often lack the necessary financial resources and experience in filing, because they are unable to judge the potential commercial applications of their invention, and because they prefer to avoid burdensome paperwork. In addition, academics are more interested in publishing than in patenting. In collaborative research projects, the titles to inventions of individual researchers are often transferred to firms in exchange for research funds.

Incentives for patenting university and NRI research results

New policy measures have recently been introduced to promote patent applications and the commercialisation of inventions patented by individual researchers. Notable examples are technology liaison offices (TLOs) and patenting support activities by government organs.

Technology Liaison Office (TLO)

The Law for the promotion of the transfer to private firms of the research results of technology in universities (*daigaku-gijutsu-iten-hou*) was enacted in 1998.

In recognition of the success of the US system and in view of the fact that huge sums of public money have been spent on university research, it was decided that Japan needed to promote the transfer of the results of university research to firms. This led to establishment of the TLOs, the organisations responsible for licensing and technology transfer of university inventions.

Although there are firms that sell or help license university patents to industry, such firms have not had much success to date in transferring technology from university to industry. The new law is designed to provide subsidies, reduce patent fees and provide other support, to organisations approved by the Ministry of Education (Monbusho) and the Ministry of Economy, Trade and Industry. It aims to provide subsidies during the start-up period of the TLOs when income cannot be expected and the expenses incurred in patent acquisition are high. Furthermore, "approved" TLOs have benefited from a reduced rate (half-rate) for patent applications and requests for examination since October 1998. As of 2000, national university researchers can be executive directors of TLOs, and can engage in technology consulting [special law for industry revitalisation (*sangyou-katsuryoku-saisei-okubetu-sochihou*)]. Furthermore, since FY 2000, the TLOs can use the premises of national universities free of charge. As of July 2000, 15 organisations had been granted TLO status.

The TLOs evaluate inventions by university researchers. If, in its evaluation, a TLO sees potential for commercialisation, it acquires the title to the invention from the inventor (a university researcher), and files a patent application. At the same time, the TLO contacts firms that might be interested in commercialising the invention and negotiates licensing agreements. Royalty revenues are paid to the university (the department and/or laboratory) and the researchers (inventors) after deduction by the TLO of expenses incurred in the handling of inventions and patents. Some TLOs adopt a firm-membership system whereby marketing to non-members is initiated after a specified lapse of time during which member firms have preferential access to the inventions and have the opportunity to indicate their interest in commercial application. If no member expresses interest, the invention will be marketed to non-member firms.

With the exception of TLOs in private universities, the TLOs are organised off the university campus, usually as private companies (investors include interested university researchers, etc.). The private universities establish their TLOs as a part of the organisation of the university. Although public subsidies are available for a limited period of time, sustaining the TLO until the invention generates sufficient revenue is a matter of concern. It is reported that, even for successful TLOs in US universities, it often takes more than ten years before revenues cover costs (Tsukamoto, 1999). Most TLOs rely on fees from member firms. However, as the number of TLOs increases, this problem will become more serious. Firms may begin to select TLOs and some TLOs may lose industry support.

Support for patenting

In 1998, the JST began to fund the acquisition of patents for inventions generated by universities and NRIs with a high potential for industrial application. Patent specialists can be called to universities and research institutes to examine the feasibility of patent applications.

With the researchers' consent, the intellectual property rights to the inventions are transferred to the JST. The JST then files applications for patents. In such cases, the JST bears the expenses for the patent application and maintenance, while the researchers' rights are preserved by contracts between the JST and the researchers. In principle, when patents are commercialised, 80% of the royalties are paid to the researchers. In the case of the NRI inventions, the patent application is jointly owned by the government and the JST. Between April and November 1998, the JST recorded 383 cases, including 86 domestic applications and ten foreign applications (JST Evaluation Committee).

The Japanese Patent Office carries out support activities related to government patents by providing consultations on patenting, searching for inventions that have patent potential, and undertaking the preparation and processing of patent applications. The Patent Office also provides various support services, such as circulating information about not-in-use patents including those of private sector firms.

2.5. *Promoting the transfer of university and NRI research results*

This section first describes the commercialisation of patents generated by research at universities and NRIs before going on to outline various measures and activities for the promotion of technology transfer.

Commercialisation of government-owned patents

As of 1998, the Japanese Government possessed 15 913 patents. Of these, 1 655 have licensing agreements (of which about two-thirds relate to the Ministry of Economy, Trade and Industry). In other words, about 10% of government-owned patents have been licensed (*Source:* STA). It should be noted, however, that this figure refers to all patents which have licensing agreements; the actual number of patents that generate royalty revenues is smaller.

The JST system

JST operates a system aimed at promoting the commercialisation of the research results of the NRIs and universities.

Risk-taking Fund for Technology Development

JST monitors promising research results generated by NRIs and universities and establishes exclusive licensing rights for the patents arising from the results. Under the "Risk-taking Fund for Technology Development" programme, JST commissions a firm to undertake the development of specific research results and provides the necessary funding. The success or failure of the development is judged by whether or not it meets the technical criteria specified in advance with a view to commercialisation. If the invention is successfully commercialised, the firm has to reimburse the funds provided by JST. In addition, the firm has to pay royalties to the JST for the use of the patents. Where development fails, the firm does not need to pay back the development costs, which are borne by JST. Thus, this programme may be seen as a conditional loan.

At the end of FY 1997, the total number of projects was 430, of which 69 were ongoing, 23 had been terminated before completion and 338 had been completed. Among the completed projects, 319 were considered successful as they met the technical criteria, while 19 were recognised as failures. Between 1966-97, a total of JPY 107 000 million was spent, of which JPY 73 400 million had been repaid by firms. Income from royalties totalled some JPY 8 000 million in the period up to 1997 and amounted to JPY 300 million in 1997 (part of which was paid to the holders of patent rights) (JST Evaluation Committee). A recent notable success was the development of the manufacturing technology for the GaN blue luminous diode based on the results of a researcher at Nagoya University.

About one-quarter of the projects recognised as successful have not resulted in commercial application. Moreover, royalty revenues have been limited. The evaluation report points out that there is room for improvement in the management of this programme, in view of the fact that firms may not be able to recover their development costs in spite of their technical "success".

Activities of the Agency for Industrial Science and Technology (AIST)

The Agency for Industrial Science and Technology (AIST) actively promotes the commercialisation of the patents in its possession. The Japanese Association for the Promotion of Industry Technology (*nihon-sangyou-gijutsu-shinkou-kyoukai*) carries out the diffusion activities for the majority of AIST patents. In October 1988, part of AIST's R&D activities was transferred to NEDO. NEDO maintains and promotes the dissemination of patents generated from its R&D. AIST possessed about 11 000 patents at March 1999, of which 1 100 were licensed and generated royalty revenues of JPY 82 million (*AIST Research Plan – 1999*).

Information dissemination

Provision of data by JST and other organisations

JST recently introduced its database (J-STORE: http://jstore.jst.go.jp) to the public in order to provide information on intellectual property rights arising from the results of R&D by universities and NRIs. Data are presented in the form of technology seeds and potential commercial applications arising from research carried out by the universities and NRIs managed by JST; they are extracted from the published literature, including patent information.

Some of the data (on researchers, research institutes, research subjects and research resources) on the R&D activities of public research institutions are provided by ReaD (the Directory Database of Research and Development Activities: http://read.jst.go.jp/). Data related to research activities in universities are provided by the National Institute for Informatics (formerly known as NACSIS) in the form of a research activity resource directory (NACSIS-DiRR). The two databases can be accessed from a single access point, the JST or NII homepages. They provide a useful source of general information on public research institutes and universities and on researchers' areas of expertise.

Although it is not a measure for the promotion of technology transfer from universities and NRIs, the Japanese version of the Bayh-Dole Act (*sangyou-katsuryoku-saisei-tokubetu-sochihou*), has recently been enacted. This law changes ownership procedures relating to the results of research carried out by private firms commissioned by the government. Before the new law, the results of research contracted by the government belonged, in principle, to the government. However, since the introduction of the new law in October 1999, the result belongs to the contract firm with the condition that it is licensed to the government free of charge when there is need for the government to use it in the public interest. This move is expected to encourage firms to commercialise the results of research commissioned by the government.

2.6. Institutional measures to promote industry, university and NRI co-operation

The following measures are intended to mitigate institutional bottlenecks in co-operation between industry, universities and NRIs.

Exchange of equipment and personnel

The national universities and NRIs are government institutions. As such, all machinery, equipment and facilities belonging to these institutions are government property, and researchers are civil servants. The conditions necessary for effective co-operation between national universities/NRIs and the private sector and/or research institutes of foreign countries may not be compatible with the rules and regulations applying to government institutions. In order to mitigate the situation and to promote co-operation, the Law for Promotion of Research Co-operation (*kenkyuu-kouryuu-sokushin-hou*), was enacted in 1986. This law has since been revised to further promote co-operation of public sector institutions with the private sector and with foreign countries.

Ownership of up to one-half of the patents generated by research commissioned by the government may be granted to firms which perform R&D under the provisions of the law and the cabinet order (Article 7 of the law). According to a 1998 revision (Article 11), land rental costs may be reduced (by up to one-half) for joint research facilities on campuses of national universities or NRIs established by firms and outside entities. At present, one instance related to Hokkaido University is in place, and one related to AIST is in the process of being implemented.

Relaxation of the rules relating to part-time employment in industry

The taking up of part-time positions in industry by researchers of national universities was formerly restricted to seven positions per person, and the total number of hours engaged was set at no more than eight hours per week. These limitations were abolished in 1997. At the same time, it became possible for researchers at national universities to engage in R&D for firms during their off-duty hours. As a result, permissions granted to researchers taking up part-time positions in industry increased (Table 12).

**Table 12. Permissions granted to national university and NRI researchers
wishing to take up part-time positions in firms**

	1996	1997	1998	1999
National universities	30 829	36 925	38 020	n.a.
NRIs	130	177	183	128

Note: 1999 = 1 April-31 December.
Source: STA (1999), *Annual Report on Science Technology Promotion in 1999.*

At the NRIs, only a small number of researchers were allowed to take up, for example, part-time teaching duties at private universities. There was no clearly specified standard, although taking up positions on a part-time basis was theoretically possible. Since 1996, METI, STA, the Ministries of Welfare, Agriculture, Posts and Telecommunications, and Transport have made it clear that researchers are allowed to take up positions on a part-time basis during their off-duty hours as long as there is no conflict of interest with their NRI positions and responsibilities. The number of researchers obtaining such permission has risen as a result (Table 12).

Employment of researchers from national universities and NRIs by venture firms

As mentioned above, researchers at national universities or NRIs are civil servants. The Japanese Constitution states that members of the civil service are mandated to serve the public. Accordingly, researchers were not allowed to take up positions in for-profit firms. To promote technology transfer from national universities and public institutions to industry, a new policy was adopted in April 2000 to allow researchers at national universities and NRIs to become directors of private firms under certain conditions.

Transforming the NRIs into independent administrative legal entities

The majority of the NRIs were set to become independent administrative legal entities in April 2001. An independent administrative legal entity (often called an "agency") is a new form of government organisation, created in conjunction with the reorganisation of government ministries and agencies initiated in 2001. A general law sets out the underlying principles of the structure and management of the independent administrative legal entity, and laws based on this general law will establish each NRI or group of NRIs.

The creation of independent administrative legal entities is designed to increase the effectiveness of government operations by assigning implementation of part of the government activities formerly performed by government organisations to these new entities. An independent administrative legal entity aims to be autonomous, responsive and transparent in its role of fostering activities of public interest that cannot be performed by private organisations. Corporate accounting systems will be introduced, funded by government. These are divided into two types: one in which a staff member has the status of a government official; the other where he/she does not. The majority of the NRIs will become entities with government official status. The government management principle for the new entities is to refrain from intervention before operations are implemented, but to perform checks and reviews after implementation. Three to five year mid-term activity plans will be drafted, proposed and approved by the minister who oversees the NRI. An annual implementation plan will be decided accordingly.

The new entities will benefit from flexibility in management not currently available to the NRIs. As contract research represents a source of revenue for the entities, the amount of contract research conducted is likely to increase. Some of the rules and regulations applying to government officials will not be applicable to researchers at the new entities, thus enabling them to respond quickly to industry demand. However, in view of the limited amount of human resources, researchers will be faced with the challenge of assigning their work hours to different activities. In this regard, the role of contract research of technology advisory services to firms *versus* the original research mission needs to be clearly stated.

In the case of national universities, studies have been performed to investigate the feasibility of transforming the national universities into independent administrative legal entities, although in a somewhat different form from the NRIs. It is expected that this reform will be implemented within the next few years. National universities and NRIs are treated as a part of the administrative organs of government, governed by the rules and regulations devised for routine administrative work. Exceptions have only been possible through the introduction of specific laws or through a flexible interpretation of the rules. It has become increasingly difficult to conduct research which necessarily involves unexpected developments and flexible management, within the rigid systems of government administrative institutions. In this context, the creation of an independent administrative legal entity is a crucial step towards increased managerial flexibility.

Organisations promoting co-operation

Several new organisations have recently been introduced with the aim of promoting co-operation between firms, universities and NRIs. These include joint research centres and research co-operation sections at national universities (Figure 6).

Figure 6. **Joint research centres and co-operation sections at national universities**

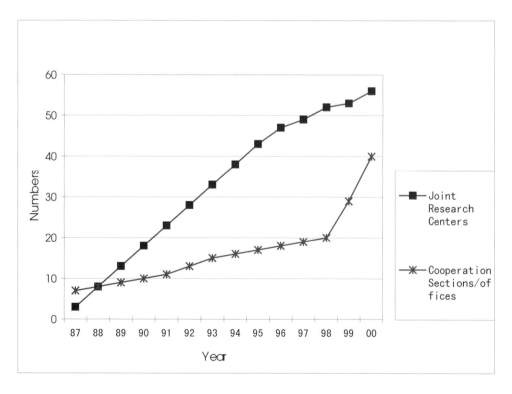

Source: Japan Scientific Monthly, Vol. 53, No. 1.

Joint research centres at national universities

Joint research centres (the names vary from one university to another) have been established at national universities. These centres aim to promote co-operation with industry and stimulate the development of technology in the regions. They provide sites where joint research with private firms and contract research can be carried out and where industry researchers are accepted. The centres serve as the window of the national universities' co-operation with industry. In addition, they are expected to provide consulting services for R&D and training for researchers.

A typical centre has its own laboratories where university professors, graduate students and researchers from private firms carry out research. Although there are a few exceptions, the centres are usually staffed by just one or two full-time personnel. University researchers are usually employed on a part-time basis. The first three centres were established in 1987. The number grew to 53 in 1999, and was planned to reach 56 in 2000.

Joint research centres play a key role in that the public and industry alike perceive them as the windows of the universities, places where industry and the regions can engage in co-operation and exchange with the universities. However, staff numbers at the centres are usually limited, and in most cases an associate professor co-ordinates the university-industry co-operation. However, university hierarchy can mean that an associate professor may find it difficult to fulfil his co-ordination role (Tsukamoto, 1999). In addition, the majority of the university researchers working at joint research centres do so on a part-time basis, thus limiting the scope of their research activities. One solution to this problem would be to increase the numbers and the rank of full-time staff at such centres.

Research co-operation sections at national universities

Research co-operation sections or offices have been increasingly established at national universities to serve as administrative windows for co-operation with industry and society. A research co-operation section or office is an organisation within a university's administrative section. Staff members are drawn from the university administration, which means that they usually lack private sector experience and have little knowledge or experience of intellectual property rights. Stringent budgetary policies imposing a reduction on civil service employees make it difficult to envisage any increase in administrative personnel. Thus, it is crucial that the sections make use of the knowledge and experience of outside experts.

NRIs

The NRIs have undertaken a similar effort to establish windows for co-operation. AIST has established co-operation promotion centres at its laboratories. Again, due to strict controls on the numbers of civil servants, most employees work part time. In other research institutes, sections for co-operation are usually established within the administrative and/or research planning sections.

2.7.   ISR between Japanese industry and foreign universities/research institutes

In 1998, JPY 119 600 million in research funds were paid by Japanese firms to other organisations. The MCA *Report on the Survey of Research and Development* divides such funding into four categories: central and local governments; special corporations; private sector; and foreign countries. Central and local governments include national universities, NRIs, public (local government operated) universities and research institutes (the private universities belong to the private sector). Related statistics are shown in Table 13.

Table 13. Research funds paid by Japanese firms to outside organisations

JPY million

	To central and local governments	To special corporations	To the private sector	To foreign countries
1992	32 785 (3.8%)	10 442 (1.2%)	736 717 (84.7%)	89 392 (10.3%)
1993	30 075 (3.5%)	10 773 (1.3%)	727 749 (85.8%)	79 490 (9.4%)
1994	35 774 (4.2%)	14 274 (1.7%)	717 906 (84.8%)	79 103 (9.3%)
1995	27 473 (3.0%)	14 952 (1.7%)	769 292 (85.0%)	93 695 (10.3%)
1996	29 158 (2.8%)	10 600 (1.0%)	876 105 (84.7%)	118 054 (11.4%)
1997	35 514 (3.1%)	9 783 (0.8%)	969 082 (83.7%)	143 291 (12.4%)
1998	49 588 (4.1%)	11 397 (1.0%)	1 000 259 (83.6%)	134 759 (11.3%)

Source: MCA Statistics Bureau, *Reports on the Survey of Research and Development.*

Expenditures to foreign countries have increased markedly (with the exception of the 1992-94 period during which R&D expenditures by the private sector firms were sluggish). The statistics do not provide details on which organisations (universities, private sector research institutes, firms, or others) in foreign countries received funding. However, it is reasonable to assume that a considerable amount went to universities in foreign countries as well as to foreign firms (see Section 3.1).

An analysis of specific sectors points to an interesting trend. The pharmaceutical sector accounts for a large share of research funding to overseas organisations (recently, a little over 40% of total overseas expenditure) (Table 14). Expenditure to central and local governments, most of which is considered as expenditure to national universities, has fallen consistently. On the other hand, expenditure to foreign countries has increased rapidly, while expenditure to the domestic private sector has remained relatively stable over the same period. "Central and local governments" and "the private sector" have seen their shares of funding decline, while the share going to "foreign countries" has dramatically increased. It should be noted that this tendency is apparent in pharmaceuticals, a sector in which R&D is close to science.

Table 14. Research funds paid by the Japanese pharmaceutical sector to outside organisations

JPY million

	To central and local governments	To special corporations	To the private sector	To foreign countries
1992	18 078 (19.7%)	1 208 (1.3%)	50 786 (55.3%)	21 820 (23.7%)
1993	16 119 (16.6%)	887 (0.9%)	52 391 (54.1%)	27 415 (28.3%)
1994	15 128 (15.6%)	747 (0.8%)	53 962 (55.7%)	26 962 (27.9%)
1995	14 601 (14.0%)	757 (0.7%)	49 747 (47.8%)	38 989 (37.5%)
1996	13 622 (11.4%)	1 681 (1.4%)	53 728 (44.8%)	50 946 (42.5%)
1997	11 777 (9.2%)	656 (0.5%)	52 986 (41.6%)	62 019 (48.7%)
1998	11 410 (9.9%)	1 027 (0.9%)	47 590 (41.1%)	55 656 (48.1%)

Source: MCA Statistics Bureau, *Reports on the Survey of Research and Development.*

A report by the Japanese Bio-industry Association highlighted the fact that firms in the bio-industry had become disenchanted with research carried out in co-operation with national universities. This partly reflects differences in industrial and university researchers' attitudes with regard to patenting and intellectual property protection. In addition, firms are concerned that co-operative research is being carried out in the absence of clearly defined contracts. This frustration may be at the root of the rise in the level of joint research with universities in foreign countries.

In addition to the pharmaceutical industry, the electric machinery and transport machinery industries typically spend large sums of research funds abroad. The electric machinery industry spent JPY 34 638 million in 1998 (JPY 31 691 million in 1997), while the transport machinery industry (including the automobile industry) spent JPY 15 489 million in 1998 (JPY 20 338 million in 1997). However, no downward trend in expenditures by these industries to central and local governments has been observed, although the share of that category in total expenditure is relatively small.

3.   Firms' perception of ISR

To better understand research co-operation between the universities, NRIs and firms, the current situation and possible areas of improvement are described below, based on qualitative data obtained from surveys of private firms.

3.1.   *Relations between universities, NRIs and large firms*

A survey of relatively large private sector firms (with capital in excess of JPY 1 000 million) carried out by the Science and Technology Agency highlighted some interesting facts about the links between universities, NRIs and private firms (STA, 1999). The following tables present different aspects of such co-operation.

Table 15. **Research co-operation with private sector firms and other organisations**

Percentages

	Domesticl firms/ same industry	Domesticl firms/ different industry	Domestic universities	NRIs	Foreign firms/ same industry	Foreign firms/ different industry	Foreign universities	Foreign research institutes	Other	No research co-operation	No response
Total industry	36.4	45.8	65.6	34.9	18.0	7.8	16.1	5.1	1.8	15.2	1.1
Manufacturing	35.0	47.1	68.1	33.7	20.5	8.8	17.3	5.4	1.8	13.1	0.9
Pharmaceuticals	71.8	41.0	94.9	48.7	71.8	7.7	48.7	15.4	2.6	0.0	0.0
Non-manufacturing	41.2	41.6	57.0	38.9	9.0	4.5	12.2	4.1	1.8	22.6	1.8
Capital:											
JPY 10 000 million to JPY 50 000 million	31.9	31.0	52.1	24.1	11.7	3.0	4.8	0.9	2.5	23.2	1.8
JPY 50 000 million to JPY 100 000 million	36.9	47.6	66.8	33.2	13.9	5.3	8.6	2.1	1.6	11.8	0.5
JPY 100 000 million to 500 000 million	38.7	58.8	78.8	43.8	23.4	9.1	25.2	8.4	1.5	8.0	5.6
JPY 500 000 million and over	50.6	75.3	88.8	64.0	40.4	32.6	59.6	21.3	0.0	5.6	0.0

Note: Number of responding firms = 986.
Source: STA (1999), *Survey of Private Firm Research Activities in 1998.*

Research co-operation

Industry research co-operation, whether with other firms, universities, NRIs or other organisations, both inside and outside of Japan, is presented in Table 16. The data are classified by manufacturing and non-manufacturing industry, and by the size of firms based on their capitalisation. The pharmaceutical industry is singled out since it performs intensive research co-operation with universities and foreign organisations. As can be seen from the table, co-operation with other organisations increases with firm size. The range of business lines becomes more diverse as firms grow, and the opportunities (and necessity) for co-operation with other firms, NRIs and universities increase.

Some two-thirds of firms co-operate with national universities and one-third co-operate with NRIs. Of firms with capital of between JPY 1 000 million and JPY 5 000 million (the smallest firms in the survey), about half co-operate with universities, and one-quarter with NRIs.

What types of co-operation do firms favour? (Table 17). Joint research, contract research and donations together constitute more than 50% of the principal forms of industry co-operation with national universities. Exchanges of research personnel tend to be more popular in larger firms, while the use of university research facilities is not significantly affected by firm size. This indicates that larger firms have more human resources and are thus more likely to send their researchers on secondment to universities.

Table 17. **Types of industry co-operation with national universities**

Percentages

	Inde-pendent joint research	Project involving several parties	Use of research facilities	Exchange of research personnel	Contract research	Donation	Other	No response
Total industry	58.0	15.8	23.8	27.2	56.7	51.8	0.3	1.2
Capital:								
JPY 5 000 million to 10 000 million	48.5	12.8	23.3	19.4	50.2	42.7	0.4	2.6
JPY 50 000 million to JPY 100 000 million	60.0	13.6	16.8	20.8	60.8	42.4	0.0	0.0
JPY 100 000 million to 500 000 million	61.6	16.7	28.2	31.0	54.6	57.9	0.5	0.9
JPY 500 000 million and over	72.2	25.3	24.1	49.4	74.7	75.9	0.0	0.0

Note: Number of responding firms = 647.
Source: STA (1999), *Survey of Private Firm Research Activities in 1998.*

What types of co-operation were used by firms that undertook co-operation with NRIs? Table 18 shows that participation in research projects involving several parties, joint research, use of university research facilities and contract research were the main forms of industry-NRI collaboration.

It would appear that firms undertake joint research with both NRIs and universities. However, industry-NRI co-operation often takes the form of research projects involving several parties, while little use is made of contract research. This is because firms tend to participate in R&D projects organised by government bodies (such as METI and other ministries or agencies), and these projects usually involve several participant firms. Firms rarely make donations to NRIs, for the reasons outlined earlier.

Table 18. **Types of research co-operation with NRIs**

Percentages

	Inde-pendent joint research	Project involving several parties	Use of research facilities	Exchange of research personnel	Contract research	Donation	Other	No response
Share of the 344 firms which reported co-operation with NRIs	43.9	46.5	39.5	29.7	38.4	3.8	0.6	2.0

Note: Number of responding firms = 344 (34.9% of all firms).
Source: STA (1999), *Survey of Private Firm Research Activities in 1998.*

For comparison purposes, different types of co-operation with universities in foreign countries are shown in Table 19. Here, joint research and contract research are the dominant modes of co-operation. Compared with industry linkages with national universities, fewer offers of donations are made to foreign universities, but exchanges of researchers with universities in foreign countries are more common. The greater importance attached to exchanges of research personnel in co-operation programmes with foreign universities is explained by the fact that staff exchanges help to mitigate the communication difficulties which are perceived as representing the most serious barrier to industry-foreign university linkages (Table 22).

Table 19. **Types of research co-operation with foreign universities**

Percentages

	Inde-pendent joint research	Project involving several parties	Use of research facilities	Exchange of research personnel	Contract research	Donation	Other	No response
Share of the 159 firms which reported co-operation with foreign universities	52.2	18.2	6.9	42.1	47.2	19.5	1.3	2.5

Note: Number of responding firms = 159.
Source: STA (1999), *Survey of Private Firm Research Activities in 1998.*

Incentives for industry co-operation

Incentives for firms to co-operate with other organisations are outlined in Table 20. Firms co-operate with national universities in order to exploit research capabilities lacking in-house, while collaboration with NRIs provides them with access to research facilities. Firms also seek to obtain research results from universities and NRIs which would not otherwise be available to them. However, firms wishing to speed up their product development and strengthen their competitiveness through the acquisition of complementary technology are more likely to co-operate with other firms than with universities or NRIs.

Table 20. **Factors motivating co-operation with domestic organisations**

Percentages

Motive	Partner		
	Other firms	Domestic universities	NRIs
Faster product development	63.9	22.6	19.2
Access to research capabilities not available in house	30.7	51.6	34.9
Access to research activities not available in house	10.4	28.6	30.2
Access to research results not available in house	29.1	47.3	41.3
Access to research methods not available in house	6.6	12.7	12.8
Access to research facilities not available in house	21.6	33.1	53.2
Strengthened competitive position through access to technology	51.8	24.0	23.3
Lower R&D expenses	24.7	8.5	15.7
Other	2.5	2.8	3.2
Number of firms reporting research links	635 (100%)	647 (100%)	344 (100%)

Note: Firms could select up to three responses.
Source: STA (1999), *Survey of Private Firm Research Activities in 1998.*

Barriers to co-operation

Between one-fifth and one-quarter of firms report no major barriers to collaboration with other firms, domestic universities or NRIs (Table 21). Although up to three responses were allowed, no single obstacle was cited by more than 50% of firms. The most notable obstacle to co-operation with NRIs stems from the issue of ownership of R&D results (cited by 42.4% of firms), while the slowness of R&D at universities was the most keenly felt obstacle to industry-university research links (35.9%). Speed was not an issue for research co-operation with foreign universities – indicating that domestic universities fail to respond to firms' expectations (Table 22). Communication difficulties are cited as posing the most serious barrier to co-operation with foreign universities.

Table 21. **Barriers to co-operation with domestic organisations**

Percentages

Barrier	Partner		
	Other firms	Domestic universities	NRIs
Difference in speed of R&D	12.0	35.9	23.5
Differences in advancement of R&D	25.7	25.7	26.2
Differences in objectives	32.9	21.3	25.0
Lack of flexibility of rules and regulations	0.9	7.0	11.0
Application of R&D results	42.7	29.2	42.4
Publication of R&D results	22.7	24.6	25.9
Others	4.9	2.8	3.2
No particular obstacles	25.5	23.5	20.1
No response	2.2	1.7	2.6
Number of firms reporting research links	635 (100%)	647 (100%).	344 (100%).

Note: Firms could select up to three responses.
Source: STA (1999), *Survey of Private Firm Research Activities in 1998.*

Table 22. **Barriers to co-operation with foreign firms and universities**

Percentages

Barrier	Partner	
	Foreign firms	Foreign universities
Issues relating to patenting, IPR	27.8	28.9
Issues relating to partner country legislation and standards	7.2	7.5
Communication difficulties	43.5	38.4
Differences in speed of R&D	13.0	13.2
Differences in advancement of R&D	29.1	21.4
Publication of R&D results	13.0	22.0
Other	5.8	3.1
No particular obstacles	11.7	19.5
No response	22.9	16.4
Number of firms reporting research links	232 (100%)	159 (100%)

Note: Firms could select up to three responses.
Source: STA (1999), *Survey of Private Firm Research Activities in 1998.*

Evaluation of research results

Table 23 summarises the responses to the question of industry's perception of the usefulness of the results of research carried out at universities and NRIs. Some firms replied that they could not evaluate the research results as they were not aware of any results. On the whole, firms judged that the quality of research results varied greatly; some are valuable, others not. More concrete complaints were that "licensing of the intellectual property generated by research results is limited, making it hard for firms to acquire the rights" and that "the speed of R&D was so slow that the results were already out-of-date".

Table 23. **Evaluation of the results of research at universities and NRIs**

Percentages

	Universities	NRIs
Research results could be evaluated	14.4	14.6
Lack of cost-effectiveness	23.1	33.3
Publication in a scientific journal is not a sufficient result	18.1	15.4
Licensing of the intellectual property generated by the research is limited, making it hard for firms to acquire the rights	29.2	20.6
The research could have been conducted in the private sector if funding had been available on the same scale	7.5	14.8
The actual research results differ significantly from the expected results	47.7	34.6
The speed of R&D was so slow that the results were already out-of-date	25.7	21.8
Lack of results which could be evaluated	3.0	3.1
Results could not be evaluated because they are not well understood	27.8	31.7
Others	3.2	3.3
No response	3.1	4.2

Note: Number of responding firms = 986. Firms could select up to three responses.
Source: STA (1999), *Survey of Private Firm Research Activities in 1998.*

Transfer of research results

In industry's view, how does the transfer of research results from universities and NRIs affect firms' R&D or corporate strategies? The surveyed firms could select a maximum of two responses: 25.4% reported that research results had no influence whatsoever; 30.2% acknowledged that they helped to "shorten the R&D and product development process"; 30.0% cited "differentiation of product from rival companies' products"; 22.9% pointed to the "creation of new business lines"; 19.0% emphasised the spin-off effects of novel ideas based on research results; finally, 12.9% used the results to "realign their research strategy".

In response to a question on the ways in which firms obtained information on research results (maximum of three answers): 79.3% of firms stated that they kept abreast of new research through "academic societies"; 57.4% through "contract and joint research"; 31.5% through "newspapers and magazines", 31.5% through personal connections within universities and/or NRIs; 21.9% through the use of "research information databases"; 11.8% through "inquiries to universities/NRIs"; 11.1% from information posted on Web sites; and 3.3% through TLOs and liaison programmes. As would be expected, academic societies were the most frequently cited source of information. Researchers generally attach great importance to presenting their findings at academic meetings and in scientific journals. For firms, it is vital to acquire information at a very early stage and this is where the academic societies play a key role.

Another often-cited means of obtaining information is through direct contact with researchers (*i.e.* through contract and joint research). Interpersonal linkages also play a key role. On the other hand, open sources such as newspapers, magazines, databases, direct inquiries and Web sites are minor sources of information. Thus, human contact seems to be an important means of acquiring information on research results.

When firms are asked to identify the main barriers to access to information (maximum of two choices), 58.3% stated that "information is not provided to the public in a way readily accessible such as in an unified database"; and 43.9% that "it is not clear where to look for research information in particular technology fields". A further 14.1% reported that the "contact point for technology transfer is not clear"; 13.0% stated that "research information posted on homepages is imperfect"; 11.4% that "in the absence of personal contacts established through joint research and other measures, it is difficult to obtain access to information". Of the remainder, 3.0% cited "other" reasons, while for 10.9% of firms, there were no obstacles.

Firms seek information that is presented in an accessible way. Rendering information user friendly requires it to be processed and modified. However, information processing takes time, with the risk of delaying the dissemination of results. Ideally, information should be presented in a user-friendly format while preserving its timeliness. An important task for firms is thus to pinpoint the sources of exploitable information among the wealth of information available.

Technology transfer

The survey examined the factors which made firms reluctant to patent the results from research by universities and NRIs (Table 24). The most often-cited impediments were: "lack of co-ordination in cases where intellectual property is shared", and that "licensing conditions are too rigid (re-licensing and exclusive rights have not been established)".

When asked which policy measures they expected from government (up to three answers were possible): 42.8% of firms requested "better dissemination of results of research undertaken in universities and NRIs"; 42.4% asked for "university and NRI results to be made available in database form"; 41.7% wanted "increased support for joint and contract research with universities and NRIs"; 33.4% requested better "promotion of venture business". A further 28.3% called for the "promotion of exchanges of personnel with universities and NRIs (acceptance by universities and NRIs of researchers from industry)"; 22.7% agreed with "promotion of systematic transfer of results through the establishment of TLOs and liaison programmes"; 16.6% were in favour of "measures to promote the circulation of unused patents".

Table 24. **Barriers to the acquisition of IPR licences from universities and NRIs**

Percentages

Barrier	
Difficulty in obtaining inventor's co-operation in examining possibilities for commercial application	21.2
Lack of co-ordination in cases where intellectual property is shared	37.1
Licensing conditions are too rigid (re-licensing and exclusive rights have not been established)	32.0
Licensing fees and royalties are too high	9.1
Contract is too vague	14.1
Licence negotiations are too time-consuming	13.5
Other	17.2
Lack of experience in acquiring licences	11.1
No response	9.1

Note: Number of responding firms = 986. Firms could select up to three responses.
Source: STA (1999), *Survey of Private Firm Research Activities in 1998.*

Industry perception of collaboration

Based on the survey results, the attitudes of large firms to university and NRI researchers may be summarised as follows:

- Firms monitor the progress of research carried out at universities and NRI in their chosen fields of interest, follow the activities of academic societies and obtain information from the media.

- If firms are familiar with the names of researchers working in areas of interest to them, they approach them directly.

- Firms then enter into co-operative relationships with researchers by setting up joint research or contract research projects. It is usually difficult to commercialise research results in the form in which they are presented at academic societies. Additional research and development is necessary. The motivation for firms to enter into co-operative relationships is to gain access to research results at the earliest stage possible. At the same time, through joint research projects firms may be able to acquire part of the intellectual property generated by the research, in the hope that they will then be able to veto licensing to other firms by refusing to relinquish their share of intellectual property.

Firms would like to see the following measures implemented:

- Easier access to research results from universities and NRIs.

190

- More user-friendly ways to monitor research activities at universities and NRIs (*e.g.* harmonised databases) and to keep abreast of the activities of academic societies.

- Policy measures to support co-operative relations with researchers at universities and NRIs. Firms' preferences are for systematic measures, such as the establishment of TLOs and liaison programmes.

- Strengthening of intellectual property rights, including the establishment of exclusive licensing rights; and co-ordination with other organisations. (In the absence of exclusive licences, other firms may use the same patents to "follow" the innovating firm. Joint research and co-ownership of research results are among the methods employed to avoid being "followed".)

3.2.   *Relationships between venture firms, universities and NRIs*

A report by NISTEP (Sakakibara *et al.*, 1999) provides valuable insight into the relations between start-ups (referred to as "venture" firms in Japan) and government research institutes/universities. A questionnaire was addressed to 2 400 venture firms listed in the NIKKEI (*Japan Economic Journal*) *Venture Business Yearbook* (1998 edition). Answers were received from 1 007 firms, giving a response rate of 42%. The *Yearbook* lists firms judged by Nihon Keizai Shimbun (NIKKEI), Inc. as having the following characteristics:

- Introducing a unique technology or know-how.

- Achieving high growth in recent years.

- Being relatively young or having recently changed its line of business.

On the other hand, no measurable criteria were applied, such as the amount of capital or numbers of employees. Although this is not a comprehensive or uniform survey, it does provide a precious source of information on the relationships between so-called venture firms and universities and NRIs. The report monitors the attitudes of venture firm managers to collaboration with national and public research institutes and universities.

Current status of research co-operation

Venture firms' interest in joint research with universities and the national/public research institutes is shown in Table 25 (the question in this survey was limited to joint research, and thus is not as comprehensive as the STA report). While about 38% of large firms co-operate with universities through joint research, about 20% of the venture firms that responded to the survey reported joint research projects with universities.

Among R&D-performing SMEs from the manufacturing sector: 24% reported co-operation with universities; 7% with NRIs; and 26% with public (local government-operated) research institutes (*White Paper on Small and Medium-sized Enterprises, 2000 edition*). Since there is a lack of consistency between "co-operation (*renkei)*" (in the SMEs) and "joint research (*kyoudou-kenkyuu)*" (for the venture survey), caution should be exercised in any comparison of these data.

Table 25. **Venture firms' joint R&D with national/public research institutes and universities**

Percentages

	With national/public research institutes	With universities
Joint R&D in progress	10.9	19.7
Joint R&D planned	5.4	7.3
No joint R&D planned, but interest shown	57.7	51.6
No joint R&D planned, no interest shown	25.8	21.3
Total	100.0	100.0

Source: Sakakibara *et al.* (1999).

Firms that had shown an interest in undertaking joint research projects but had yet to enter the planning stage were asked why they had not yet done so (Table 26).

Table 26. **Reasons why co-operation has not progressed**

Percentages

	With national/public research institutes	With universities
Unaware of the existence of national/public research institutes (universities)	19.5	18.3
Unaware of the types of research activity carried out at each of the research institutes (university laboratories)	24.3	30.6
Unfamiliar with the procedures for entering into joint R&D collaboration	11.7	9.3
Procedures for entering into joint R&D are complicated	4.4	1.8
Problems with regard to the commercial application/ownership of results of joint R&D	8.8	5.8
No appropriate topic for joint R&D	26.4	27.7
Other	4.6	6.0
Total	100.0	100.0

Source: Sakakibara *et al.* (1999).

Clearly, if there is no appropriate topic for joint R&D, collaboration is unlikely to take place. However, the major reason cited by venture firms for not undertaking joint research co-operation is lack of information on the types of national/public research institutes and universities that exist and on their research activities. Venture firms are more reluctant to collaborate with national/public research institutes than with universities because of the more complex procedures relating to co-operation and to the allocation of research results.

Transfer of research results

Venture firms were asked to describe their experience in using patents generated by universities and national/public research institutes for their business (Table 27). Surprisingly, less than 10% of firms made use of patents owned by universities and NRIs, although more than 60% of firms had expressed an interest. On the other hand, some 30% of venture firms reported that they had no experience and no interest in the research results of universities and national/public research institutes.

Table 27. **Venture firms' use of patents owned by national/public research institutes and universities**

Percentages

	With national/public research institutes	With universities
Patents have been used	8.3	9.5
Patents have never been used (although the firm is interested)	62.7	61.9
Patents have never been used (the firm is not interested)	28.8	28.4
Total	100.0	100.0

Source: Sakakibara *et al.* (1999).

When firms were asked why their interest in patent use does not lead to a licensing contract (Table 28), the main reason was their lack of knowledge about available patents.

Table 28. **Why interest in patent use does not lead to licensing**

Percentages

	With national/public research institutes	With universities
Lack of information on what kinds of patents are available	65.4	69.0
Lack of familiarity with procedures for patent use	11.0	8.3
Procedures for patent use are complicated	4.9	3.8
Royalties are too high	4.2	4.3
Other	14.2	14.3
Total	100.0	100.0

Source: Sakakibara *et al.* (1999).

The data indicate that venture firms have difficulty in obtaining information about the research activities and results of national/public research institutes and universities, and especially about patents owned by them. Therefore, if firms had better knowledge about patents owned by the national/public research institutes and universities, it is likely that their use would increase. Although some patents may be found to be of little use to venture firms, the fact that currently about 10% of firms use patents points to an untapped potential. Increasing the awareness of venture firms of the research activities and research results of the national/public research institutes and universities is key to strengthening co-operative ties between the public and private sectors.

4.   Concluding remarks

Japan's higher education system may not have as good a reputation as its primary education, which is widely praised for having achieved universal education around the turn of the century. However, a close look at the history of Japan's economic development shows the significant contribution of universities to economic growth. The universities produced highly educated scientists and engineers, and helped industry to acquire and absorb advanced technology from the West (Odagiri and Goto, 1996).

This does not mean that industry-science relationships in Japan today are without problems. With the possible exception of those in engineering faculties, university researchers are generally reluctant to collaborate with industry. Academics are evaluated by their research achievements, not by their contribution to industry. In addition, the major universities are national and are therefore subject to strict government regulations, rendering close co-operation with industry difficult. This situation has led to the development of informal relationships between university professors from engineering departments and firms.

Firms have an interest in developing and maintaining good contacts with professors at elite universities. First, they are in a good position to hire promising students and professors are often influential in their students' career decisions. Second, especially in the past when firms' research capacities were less well-developed, professors often had advance knowledge about science and technological developments, both at home and in the United States and Europe. This knowledge helped firms to introduce new products and processes and solve technical problems. Third, industrial researchers sent to university laboratories were able to obtain advanced training and sometimes advanced degrees. Fourth, firms were interested in the technology being developed at the universities. By maintaining a good relationship, a company could develop products or processes based on the technology developed by a professor on an exclusive basis. The professors benefited too, as the firms provided funding. In addition, professors were able to make use of industrial researchers (who remained on the company payroll). Technological developments in industry were also of great interest to professors in deciding their own directions of research.

These informal relationships between professors and firms were mutually advantageous, and may have played a substantive role in Japan's industry-science linkages. Research funding was provided, often based not on contract research with specified rights and obligations but on personal, tacit agreements between professors and firms. The professor published the research findings and the intellectual property was usually given free of charge to the industrial user.

However, such arrangements were informal and lacked transparency. In addition, large established firms with long-standing relationships with universities benefited from the system, but it was more difficult for newer and smaller firms to approach the universities. There have also been cases where the lack of transparency of flows of funds from industry to professors has been criticised by society. Given the importance of co-operation between universities and industry, clear rules regarding access to university resources and know-how are crucial. Openness is also required in promoting spin-offs by university professors, supporting start-ups by graduates, commercialising research results by university professors, and other forms of industry-science relations.

A series of recent government initiatives reflects a desire to address these problems by creating more open and transparent industry-science relationships as described in this chapter. It is too early to assess the impact of these initiatives, but it is certainly crucial to design a new template for ISRs, bearing in mind their complexity. Particular attention needs to be paid to the following points.

First, the industry-science relationship must be seen as a process of knowledge creation. The knowledge created in universities has two main uses for industry: it can provide firms with the know-how required to undertake new R&D projects; and it can help industry researchers solve the problems encountered in their R&D activities. Helping to solve the problems which arise during the innovation process is possibly the most important role played by the universities as the majority of new R&D projects are implemented in response to demand by customers and industry. This point should be taken into account in any discussion on policies for the promotion of university patenting and university spin-offs. The primary channel through which firms obtain information on university research is articles published by university researchers in academic journals. Therefore, in the interest of stimulating innovation, it is extremely important to maintain active research at the universities and ensure speedy and free publication of research results. Care should be taken not to impede this critical role in the promotion of other goals such as university patenting, university spin-offs, and increased financial support by industry. Policies must be carefully designed to promote both these goals and not to promote one at the expense of the other.

Second, since the industry-science linkage is a part of the knowledge creation process, it is important that the recipients of this knowledge – firms – commit themselves to this process. The recent literature on technology transfer emphasises that a deliberate learning effort on the part of the recipient is the key to successful technology transfer. In this regard, Japanese firms, particularly large firms, have been reluctant to commit themselves. Industry's major concern, some would say its only concern, has been to maintain amicable relationships with professors at the leading universities so that they have privileged access to a supply of recruits. As they grew in size and wealth, large Japanese firms with well-established central research laboratories employing large numbers of top-level researchers had little need for research co-operation with the universities. In the late 1980s, a number of large firms set up basic research laboratories outside their central research laboratories; these units became increasingly self-sufficient and self-contained. Of course, this situation might be a reflection of firms' dissatisfaction with the Japanese universities, but it is also true that commitment by firms is required if the domestic university system is to improve. With the economic downturn, firms are finding it increasingly difficult to do everything on their own, and are having to learn to work closely with universities.

The national laboratories are divided into two types, according to their mission. The first type is related to research traditionally attributed to the government (agriculture, defence, space, public health). The second type of laboratory aims to help industries develop and acquire technology. This dual role exists in many other countries. In Japan, contrary to popular belief, the majority of government labs belongs to the former category, with government labs related to agriculture being the largest. This remains true today, although agriculture is now only a minor industry. The majority of the second type of government labs came under METI's Agency of Industrial Science and Technology, they worked closely with industry and played an important role in the take-off of key Japanese industries such as electronics and computers. However, they have gone through a series of changes in recent decades. In the 1980s, at the height of the trade dispute with the United States and others, Japan was criticised for free-riding basic research conducted by other countries. In addition, it was argued that basic research became more important for industry as it approached the technology frontier. At the same time, Japanese firms grew in size and resources and were able to acquire technological capabilities of their own. Against this background, the METI labs began to emphasise basic research or science. Researchers welcomed this trend but it made the relationship of these labs to industry more remote. The focus of these labs became blurred and they increasingly resembled the universities. Again, with the downturn of the economy and the effort to redefine their mission more clearly, the labs under METI'S AIST were to be merged into a single lab in 2001, with an emphasis on close collaboration with industry. However, given the huge increase in the technological capabilities of industry, how the government labs can help to promote industrial technology today remains to be seen. Clearly, there is a need for more research into the role of government labs in the national innovation systems of developed countries.

The role that universities and national research institutes are expected to play in the Japanese innovation system has become more critical as the importance of science to industrial technology has grown. This is true not only in high-technology sectors such as electronics or biotechnology, but for industry in general. The relationships which exist between technology and science, and between industry and university/government labs, are complex. With the growing interlinkages between rapid advances in technology and science, on the one hand, and institutions and policies, on the other, we are currently witnessing the birth of a new Japanese innovation system.

NOTES

1. *Source:* STA. It should be noted that in 1998 a single research institute attached to the national hospital system carried out a large value of contract research to the sum of JPY 4 100 million. This was made possible through the existence of a special account set up for the national hospital system.

2. Calculated by authors from the *AIST Research Plan – 1999*. Since the number of counterpart organisations is calculated for each joint research project, the same organisation may be counted more than once.

REFERENCES

Agency of Industrial Science and Technology, Ministry of International Trade and Industry (1999), *AIST Research Plan – 1999* (in Japanese), Japanese Industry Technology Promotion Association.

Goto, Akira and Nagata Akiya (1999), "Role of Universities in Innovation System in Japan: Knowledge Flow between Industry and University" (in Japanese), *Working Paper WP#99-07*, Institute of Innovation Research, Hitotsubashi University, Japan.

Institute of Future Technology (1999), "Report on the Dynamic Research System and Researchers' Lifecycle" (in Japanese), Japan.

Japan Bio-industry Association (1999), "Research of Universities for the Promotion of Bio-industry in Japan" (in Japanese), Japan.

Japan Science and Technology Corporation Evaluation Committee (JST) (1999), "Report on Evaluation of Technology Transfer Activities of the JST" (in Japanese).

Japan Society for the Promotion of Sciences (2000), "Feature: New Development in Industry-University Co-operation: Related Budget and Laws", *Japan Scientific Monthly (Gakujutu Geppou)*, Vol. 53, No. 1, pp.70-80.

Kusunoki, Ken (1997), "ERATO: The Innovation in Basic Research System" (in Japanese), *Working Paper WP#97-11*, Institute of Innovation Research, Hitotsubashi University, Japan.

Management and Coordination Agency Statistics Office (annual), *Report on the Survey of Research and Development* (in Japanese), Japan Statistics Society.

Ministry of Education (annual), *Education Policy in Japan* (in Japanese), Printing Bureau of Ministry of Finance, Japan.

Ministry of Education (1999), *School Basic Survey – 1999* (in Japanese), Printing Bureau of Ministry of Finance, Japan.

Ministry of Education (1999), *Statistical Abstract of Education, Science, Sports and Culture – 1999* (in Japanese), Printing Bureau of Ministry of Finance, Japan.

Narin, Francis, Kimberly S. Hamilton and Dominic Olivastro (1997), "The Increasing Linkage Between US Technology and Public Science", *Research Policy*, Vol. 26, pp. 317-330.

National Institute of Science and Technology Policy (2000), *Science and Technology (S&T) Indicators (2000 edition)* (in Japanese), NISTEP Report No. 66.

Nishikata, Chiaki (1992), "The Condition for the Increase in Science and Engineering Doctors in Japan" (in Japanese), NISTEP Research Material No. 24.

Odagiri, Hiroyuki and Akira Goto (1996), *Technology and Industrial Development in Japan: Building Capabilities by Learning Innovation, and Public Policy,* Oxford University Press.

Rogers, Steven (2000), "Technology Transfer in Japan: A View from Outside", *Business Review*, Vol. 47, No. 3, January.

Sakakibara, Kiyonori, Kazunori Kondo, Noboru Maeda, Shigeru Tanaka, Yoshihisa Koga and Hiroyuki Ayano (1999), "Research on Venture Business and Entrepreneurs in Japan, 1998" (in Japanese), NISTEP Report No. 61.

Sakakibara, Kiyonori (2000), "Industry-University Cooperation and Knowledge Production System" (in Japanese), *Organizational Science (Sosiki Kagaku)*, Vol. 34, No. 1, pp. 45-53.

Sienko, T. (1997), "Comparison of Japanese and U.S. Graduate Programs in Science and Engineering", National Institute of Science and Technology Policy, Discussion Paper No. 3.

Science and Technology Agency (STA) (1999), "The Survey of Private Firm Research Activities in 1998" *(Minkan Kigyo-no Kenkyuu Katsudou-ni kansuru Chousa)* (in Japanese), Japan.

Science and Technology Agency (STA) (1999), *Annual Report on Science and Technology Promotion in 1999*, June, Japan.

Tsukamoto, Yoshiaki (1999), "Study on the Academia-Industry Co-operation Systems in Research Universities" (in Japanese), *Journal of Science Policy and Research Management (Kenkyuu Gijutsu Keikaku)*, Vol. 14, No. 3, pp. 190-204.

Tsukamoto, Yoshiaki and Koji Nishio (1999), "Considerations on Improving the Industry-University Co-operation System in Japan" (in Japanese), *Abstract of Presentations at the Annual Meeting of Japan Society for Science Policy and Research Management*, November.

Yamamoto, S. (1997), "The Role of the Japanese Higher Education System in Relation to Industry", in Goto and Odagiri (eds.), *Innovation in Japan*, Oxford University Press, pp. 294-307.

Zucker, L. and M. Darby (1998), "Capturing Technological Opportunity via Japan's Star Scientists: Evidence from Japanese Firms' Biotechnology Patents and Products", *NBER Working Paper* No. W6360, Cambridge, Mass.

OECD PUBLICATIONS, 2, rue André-Pascal, 75775 PARIS CEDEX 16
PRINTED IN FRANCE
(92 2002 05 1 P) ISBN 92-64-19741-9 – No. 52411 2002